...tion

CHEMISTRY
SUPER REVIEW®

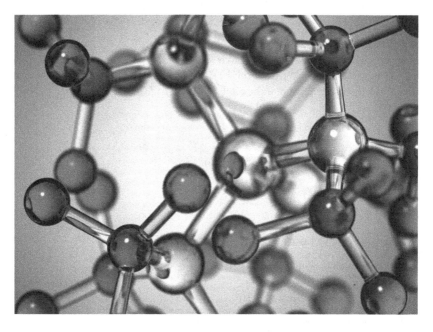

By the Staff of
Research & Education Association

Research & Education Association
Visit our website at: www.rea.com

Research & Education Association
61 Ethel Road West
Piscataway, New Jersey 08854
E-mail: info@rea.com

CHEMISTRY
SUPER REVIEW®

Published 2015
Copyright © 2013 by Research & Education Association, Inc.
Prior edition copyright © 2000 by Research & Education Association, Inc. All rights reserved. No part of this book may be reproduced in any form without permission of the publisher.

Printed in the United States of America

Library of Congress Control Number 2012949164

ISBN-13: 978-0-7386-1116-7
ISBN-10: 0-7386-1116-6

SUPER REVIEW® and REA® are registered trademarks of Research & Education Association, Inc.

REA's *Chemistry Super Review*®

Need help with Chemistry? Want a quick review or refresher for class? This is the book for you!

REA's *Chemistry Super Review*® gives you everything you need to know!

This *Super Review*® can be used as a supplement to your high school or college textbook, or as a handy guide for anyone who needs a fast review of the subject.

- **Comprehensive, yet concise coverage** – review covers the material that is typically taught in a beginning-level chemistry course. Each topic is presented in a clear and easy-to-understand format that makes learning easier.

- **Questions and answers for each topic** – let you practice what you've learned and build your chemistry skills.

- **End-of-chapter quizzes** – gauge your understanding of the important information you need to know, so you'll be ready for any chemistry problem you encounter on your next quiz or test.

Whether you need a quick refresher on the subject, or are prepping for your next test, we think you'll agree that REA's *Super Review*® provides all you need to know!

Available Super Review® Titles

ARTS/HUMANITIES
Basic Music
Classical Mythology
History of Architecture
History of Greek Art

BUSINESS
Accounting
Macroeconomics
Microeconomics

COMPUTER SCIENCE
C++
Java

HISTORY
Canadian History
European History
United States History

LANGUAGES
English
French
French Verbs
Italian
Japanese for Beginners
Japanese Verbs
Latin
Spanish

MATHEMATICS
Algebra & Trigonometry
Basic Math & Pre-Algebra
Calculus
Geometry
Linear Algebra
Pre-Calculus
Statistics

SCIENCES
Anatomy & Physiology
Biology
Chemistry
Entomology
Geology
Microbiology
Organic Chemistry I & II
Physics

SOCIAL SCIENCES
Psychology I & II
Sociology

WRITING
College & University Writing

About Research & Education Association

Founded in 1959, Research & Education Association is dedicated to publishing the finest and most effective educational materials—including study guides and test preps—for students in middle school, high school, college, graduate school, and beyond.

Today, REA's wide-ranging catalog is a leading resource for teachers, students, and professionals. Visit *www.rea.com* to see a complete listing of all our titles.

Acknowledgements

We would like to thank Pam Weston, Publisher, for setting the quality standards for production integrity and managing the publication to completion; Larry B. Kling, Vice President, Editorial, for his supervision of revisions and overall direction; Kelli Wilkins, Copywriter, for coordinating development of this edition; Transcend Creative Services, for typesetting this edition; and Christine Saul, Senior Graphic Designer, for designing our cover.

Contents

CHAPTER 1

Introduction

1.1 Matter and Its Properties

1.1.1 Definition of Matter

Matter occupies space and possesses mass. Mass is an intrinsic property of matter.

Weight is the force, due to gravity, with which an object is attracted to the earth.

Force and mass are related to each other by Newton's equation (Newton's Law), $F = ma$, where F = force, m = mass, and a = acceleration. Weight and mass are related by the equation $w = mg$, where w = weight, m = mass, and g = acceleration due to gravity.

Note that the terms "mass" and "weight" are often (incorrectly) used interchangeably in most literature.

1.1.2 States of Matter

Matter occurs in three states or phases: solid, liquid, and gas. A solid has both a definite size and shape. A liquid has a definite volume but takes the shape of the container, and a gas has neither definite shape nor definite volume.

1.1.3 Composition of Matter

Matter is divided into two categories: distinct substances and mixtures. Distinct substances are either elements or compounds. An element is made up of only one kind of atom. A compound is composed of two or more kinds of atoms joined together in a definite composition.

Mixtures contain two or more distinct substances more or less intimately jumbled together. A mixture has no unique set of properties: it possesses the properties of the substances of which it is composed.

In a homogeneous mixture, the composition and physical properties are uniform throughout. Only a single phase is present. A homogeneous mixture can be gaseous, liquid, or solid. A heterogeneous mixture, such as oil and water, is not uniform and consists of two or more phases.

1.1.4 Properties of Matter

Extensive properties, such as mass and volume, depend on the size of the sample. Intensive properties, such as melting point, boiling point, and density, are independent of sample size.

Physical properties of matter are those properties that can be observed, usually with our senses. Examples of physical properties are physical state, color, and melting point.

Chemical properties of a substance are observed only in chemical reactions involving that substance.

Reactivity is a chemical property that refers to the tendency of a substance to undergo a particular chemical reaction.

Chemical changes are those that involve the breaking and/or forming of chemical bonds, as in a chemical reaction.

Physical changes do not result in the formation of new substances. Changes in state are physical changes.

1.2 Conservation of Matter

1.2.1 Law of Conservation of Matter

In a chemical change, matter is neither created nor destroyed, but only changed from one form to another. This law requires that "material balance" be maintained in chemical equations.

1.3 Laws of Definite and Multiple Proportions

1.3.1 Law of Definite Proportions

A pure compound is always composed of the same elements combined in a definite proportion by mass.

Problem Solving Example:

Q It has been determined experimentally that two elements, A and B, react chemically to produce a compound or compounds. Experimental data obtained on combining proportions of the elements are:

	Grams of A	Grams of B	Grams of Product Compound
Experiment 1	6.08	4.00	10.08
Experiment 2	18.24	12.00	30.24
Experiment 3	3.04	2.00	5.04

(a) Which two laws of chemical change are illustrated by the above data? (b) If 80 g of element B combines with 355 g of a third element C, what weight of A will combine with 71 g of element C? (c) If element B is oxygen, what is the equivalent weight of element C?

 (a) If one adds the weight of A to the weight of B and obtains the weight of the compound formed, the law of conservation of matter is illustrated. This law states that there is no detectable gain or loss of matter in a chemical change. Using the data from the experiments described, you find the following:

For experiment 1   6.08 g A + 4.00 g B = 10.08 g of compound
     6.08 + 4.00 = 10.08

For experiment 2   18.24 g A + 12.00 g B = 30.24 g of compound
     18.24 + 12.00 = 30.24

For experiment 3   3.04 g A + 2.00 g B = 5.04 g of compound
     3.04 + 2.00 = 5.04

From these calculations one can see that the law of conservation of matter is shown.

Another important law of chemistry is the law of definite proportions. This law is stated: when elements combine to form a given compound, they do so in a fixed and invariable ratio by mass. One can check to see if this law is adhered to by calculating the ratio of the weight of A to the weight of B in the three experiments. If all of these ratios are equal, the law of definite proportions is shown.

$$\exp 1 \frac{A}{B} = \frac{6.08\,\text{g A}}{4.00\,\text{g B}} = 1.52$$

$$\exp 2 \frac{A}{B} = \frac{18.24\,\text{g A}}{12.00\,\text{g B}} = 1.52$$

$$\exp 3 \frac{A}{B} = \frac{3.04\,\text{g A}}{2.00\,\text{g B}} = 1.52$$

The law of definite proportions is illustrated here.

(b) From the law of definite proportions, one can find the number of grams of B that will combine with 71 g of C. After this weight is found, one can find the number of grams of A that will react with 71 g of C by finding the amount of A that reacts with that amount of B. It is assumed that the amount of A that reacts with 71 g of C is equal to the amount of A that will react with the amount of B that reacts with 71 g of C. The amount of A that will react with this amount of B can be found by remembering, from the previous section of this problem, that A reacts with B in a ratio of 1:1.5.

(1) Finding the amount of B that would react with 71 g of C.

One is told that 80 g of B reacts with 355 g of C. By the law of definite proportions, a ratio can be set up to calculate the number of grams of B that will react with 71 g of C.

Let x = the number of grams of B that will react with 71 g of C.

$$\frac{80\,\text{g B}}{355\,\text{g C}} = \frac{x\,\text{g B}}{71\,\text{g C}}$$

$$x = \frac{71\,\text{g} \times 80\,\text{g}}{355\,\text{g}} = 16\,\text{g}$$

16 g of B will react with 71 g of C.

(2) It is assumed that the same amount of A that will react with 16 g of B will react with 71 g of C. Therefore, using the fact that the ratio of the amount of A that reacts to the amount of B is equal to 1.5 (this fact was obtained in part (1)), one can calculate the amount of A that will react with 71 g of C.

Let x = the number of grams of A that will react with 16 g of B.

$$\frac{x\text{g A}}{16\text{g B}} = 1.5$$

$$x = 16\text{ g} \times 1.5 = 24\text{ g}$$

24 g of A will react with 16 g of B or 71 g of C.

(c) In finding the equivalent weight of C when B is taken to be oxygen, the law of definite proportions is used again. The equivalent weight of oxygen is 8 g. Knowing that 16 g of B reacts with 71 g of C, one can set up the following ratio:

Let x = weight of C if the weight of B is taken to be 8 g.

$$\frac{71\text{g C}}{16\text{g B}} = \frac{x\text{g C}}{8\text{g B}}$$

$$x = \frac{8\text{g} \times 71\text{g}}{16\text{g}} = 35.5\text{g}$$

The equivalent weight of C when B is taken to be oxygen is 35.5 g.

1.3.2   Law of Multiple Proportions

When two elements combine to form more than one compound, different masses of one element combine with a fixed mass of the other element such that those different masses of the first element are in small whole-number ratios to each other.

Problem Solving Example:

Q Two different compounds of elements A and B were found to have the following composition: first compound, 1.188 g of A combined with 0.711 g of B; second compound, 0.396 g of A combined with 0.479 g of B. Show that these data are in accordance with the law of multiple proportions.

A The law of multiple proportions can be stated: when two elements combine to form more than one compound, the different masses of one that combine with a fixed mass of the other are in the ratio of small whole numbers. This means that if one solves for the expected amount of B that is used in forming the second compound from the ratio of A:B in experiment 1, the experimental amount should be a multiple of the calculated value. This is seen more clearly after looking at the data.

In experiment 1, A combines with B in a ratio of 1.188 g A:0.711g B, or 1:0.598. In experiment 2, A combines with B in a ratio of 0.396 A:0.479 B, or 1:1.21. The law of multiple proportions states that 0.598 should be a small multiple of 1.21. Thus, 1.21/0.598 = 2 and the law is supported.

1.4   Energy and Conservation of Energy

1.4.1   Definition of Energy

Energy is usually defined as the ability to do work or transfer heat.

1.4.2   Forms of Energy

Energy appears in a variety of forms, such as light, sound, heat, mechanical energy, electrical energy, and chemical energy. Energy can be converted from one form to another.

Two general classifications of energy are potential energy and kinetic energy. Potential energy is due to position in a field. Kinetic energy is the energy of motion and is equal to one-half of an object's mass, m, multiplied by its speed, s, squared: $KE = \dfrac{1}{2} ms^2$.

1.4.3   Law of Conservation of Energy

Energy can be neither created nor destroyed, but only changed from one form to another.

1.5   Measurement

Numbers that arise as the result of measurement may contain zero or more significant figures (significant digits). The general rule for multiplication or division is that the product or quotient should not possess any more significant figures than does the least precisely known factor in the calculation.

For addition and subtraction, the rule is that the absolute uncertainty in a sum or difference cannot be smaller than the largest absolute uncertainty in any of the terms in the calculation, i.e., the number of significant figures is limited by the number of digits to the right of the decimal point in the term that is known to the fewest decimal places.

Some examples of assigning significant figures (SF):

Quantity	No. of SF's
0.006110	4 (6.110×10^{-3})
7,685,000	4 (ambiguous)
7.685×10^6	4 (ambiguity removed by use of scientific notation)
1.2×10^8	2

Examples of arithmetic using the rules for significant figures:

$$(3.39 \times 10^{-3}) \div (3.0 \times 10^{0}) = (1.1 \times 10^{-3})$$

3 SF		2 SF		2 SF
1.4963	+	0.01	=	1.51
4 decimal		2 decimal		2 decimal
places		places		places

1.5.1 Length

1 meter = 39.37 inches, 12 inches = 1 foot

1 meter = 100 cm, 1 meter = 1,000 mm, 1 inch = 2.54 cm

1.5.2 Volume

Volume = height $\times$ length $\times$ width [for a rectangular solid];

$\dfrac{4}{3} \pi r^3$ [sphere]; $\pi r^2 \bullet$ height [right circular cylinder]

1 L = 1,000 cubic centimeters (cc), 1 inch^3 = 16.4 cc

1.5.3 Mass

1 g = 1,000 mg, 1 kg = 1,000 g

1.5.4 Density

$$\text{density} = \frac{\text{mass}}{\text{volume}} \left[\frac{\text{g}}{\text{mL}} \right]$$

Problem Solving Example:

 Calculate the density of a block of wood that has a mass of 750 kg and has the dimensions 25 cm × 0.1 m × 50 m.

 Density is a measure of mass per unit volume and is usually expressed in g/mL. Therefore, one must find the mass of this block in grams and the volume in milliliters. The density is then found by dividing the mass by the volume.

1 kg = 1,000 g; therefore, 750 kg = 750 × 1,000 g = 7.5×10^5 g. Since 1 mL = 1 cm^3, to find the volume in milliliters, all of the dimensions must be converted to centimeters first.

1 m = 100 cm; thus, 0.1 m = 10 cm and 50 m = 5,000 cm.

Volume = 25 cm × 10 cm × 5,000 cm = 1.25×10^6 mL.

Solving for the density:

$$\text{density} = \frac{\text{mass}}{\text{volume}} = \frac{7.5 \times 10^5 \text{ g}}{1.25 \times 10^6 \text{ mL}} = 0.6 \, \text{g/ mL}$$

1.5.5   Temperature Scales

°C = 5/9(°F – 32°) Celsius   °F = 9/5(°C) + 32° Fahrenheit

K = °C + 273.15 Kelvin   °R = °F + 459.67 = 9/5(K) Rankine

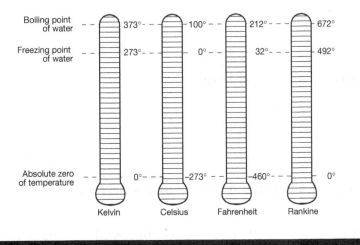

	Kelvin	Celsius	Fahrenheit	Rankine
Boiling point of water	373°	100°	212°	672°
Freezing point of water	273°	0°	32°	492°
Absolute zero of temperature	0°	–273°	–460°	0°

Problem Solving Example:

 The freezing point of silver is 960.8 °C and the freezing point of gold is 1,063.0 °C. Convert these two readings to Kelvin (K), Fahrenheit (°F), and Rankine (°R).

A **Kelvin**: Temperatures measured in Celsius (°C) are converted to K by adding 273.15 to the original measurement.

freezing point of silver = 960.8 °C + 273.15 = 1,234.0 K

freezing point of gold = 1,063.0 °C + 273.15 = 1,336.2 K

Fahrenheit: °C are converted to °F by using the equation °F = 9/5(°C) + 32.

freezing point of silver = 9/5(960.8 °C) + 32 = 1,761 °F

freezing point of gold = 9/5(1,063.0 °C) + 32 = 1,945 °F

Rankine: The Rankine scale is an absolute scale used by engineers. Its unit is the Fahrenheit degree. Absolute zero is equal to zero degrees Rankine. Convert K to °R by using the equation

$$°R = 9/5(K).$$

freezing point of silver = 9/5(1,234.0 K) = 2,221.2 °R

freezing point of gold = 9/5(1,336.2 K) = 2,405.2 °R

1.5.6 Exponential Notation

$3,630,000 = 3.630000 \times 10^6$,

$0.000000123 = 1.23 \times 10^{-7}$

1.5.7 Multiplication (Exponents Are Added)

Example: $(5.1 \times 10^{-6})(2 \times 10^{-3}) = 10.2 \times 10^{-9} = 1.02 \times 10^{-8}$
$$= 1 \times 10^{-8} \text{ (1 SF)}$$

1.5.8 Division (Exponents Are Subtracted)

Example: $(1.5 \times 10^3) \div (5.0 \times 10^{-2}) = 0.30 \times 10^5 = 3.0 \times 10^4 \text{ (2 SF)}$

1.5.9 Factor Label Method of Conversion

Dimensional quantities ("units") are treated as algebraic quantities in expressions.

Example:

$$1 \times 10^{-3} \text{ kilogram} \times \frac{1 \times 10^{-3} \text{ grams}}{1 \text{ kilogram}} \times \frac{1 \times 10^{-3} \text{ milligrams}}{1 \text{ gram}} = 1 \times 10^3 \text{ mg}$$

Quiz: Introduction

1. A box of baking soda is best described as

    (A) an element.

    (B) a compound.

    (C) a homogeneous mixture.

    (D) a heterogeneous mixture.

    (E) an aggregate of homogeneous mixtures.

2. When elements combine to form a given compound, they do so in a fixed and invariable ratio by

    (A) weight.

    (B) proportion.

    (C) mass.

    (D) state.

    (E) volume.

3. Which of the following is a chemical property?

    (A) Melting point   (D) Mass

    (B) Density   (E) Burning

    (C) Viscosity

4. Which of the following is incorrect?

    (A) 1 liter = 1,000 cm^3

    (B) 1 meter = 100 cm

    (C) 1 milliliter = 1 cm^3

    (D) 1 liter = 1 meter^3

    (E) 1 milliliter = 10^{-6} meter^3

5. Which of the following is not a homogeneous mixture?

    (A) Sugar in water   (D) Gasoline

    (B) Salt in water   (E) Soft drinks

    (C) Sand in water

6. A Fahrenheit temperature is converted to a corresponding Centigrade (Celsius) value by the equation: $°C = \dfrac{5}{9}(°F - 32)$. A recorded Fahrenheit value is 82.0°. Its value on the Kelvin temperature scale is approximately

    (A) 355.0 K.   (D) 246.0 K.

    (B) 27.7 K.   (E) 300.9 K.

    (C) 83.7 K.

7. The law of conservation of matter states that matter is conserved in terms of

    (A) pressure.   (D) mass.

    (B) volume.   (E) density.

    (C) temperature.

8. A comparison of the weights of equal volumes of carbon monoxide and carbon dioxide illustrates

    (A) the law of definite composition.

    (B) the law of multiple proportions.

    (C) conservation of matter.

    (D) conservation of energy.

    (E) conservation of matter and energy.

9. The mass of an object will change when this changes:

    (A) the force of gravity   (D) acceleration

    (B) volume   (E) None of the above.

    (C) density

10. The quantity 2.4×10^{-5} is equal to

    (A) 2.4×10^2 kg.   (D) 2.4×10^{-1} kg.

    (B) 2.4×10^{-11} kg.   (E) None of the above.

    (C) 2.4×10^8 kg.

ANSWER KEY

1. (B)   6. (E)
2. (C)   7. (D)
3. (E)   8. (B)
4. (D)   9. (E)
5. (C)   10. (B)

CHAPTER **2**

Stoichiometry,

Chemical Arithmetic

2.1   The Mole

One mole of any substance is that amount that contains Avogadro's number of particles (atoms, molecules, ions, electrons). Avogadro's number is (approximately) 6.02×10^{23}.

$$\text{number of moles} = \frac{\text{mass}}{\text{molecular weight}}$$

Note the dimensional analysis:

$$\text{mole} = \frac{\text{grams}}{(\text{grams} / \text{mole})}.$$

2.2   Atomic Weight

The gram-atomic weight of any element is defined as the mass, in grams, that contains 1 mole of atoms of that element.

For example, approximately 12.0 g of carbon, 16.0 g of oxygen, and 32.1 g of sulfur each contain 1 mole of atoms.

(This refers to "average atomic weight"; see also "isotopes," page 49.)

Problem Solving Example:

 During the course of World War I, 1.1×10^8 kg of poison gas was fired on Allied soldiers by German troops. If the gas is assumed to be phosgene ($COCl_2$), how many molecules of gas does this correspond to?

 To solve this problem, we must convert mass to number of moles and then multiply by Avogadro's number to obtain the corresponding number of molecules.

The number of moles of gas is given by

$$moles = \frac{mass\ (grams)}{molecular\ weight\ of\ COCl_2\ (g\ /\ mole)}.$$

The mass of gas is $1.1 \times 10^8 = [1.1 \times 10^8\ kg] \times [1.0 \times 10^3\ g/kg] = 1.1 \times 10^{11}$ g. The molecular weight of $COCl_2$ is obtained by adding the atomic weights (atm. wt.) of its constituents. Thus,

$$
\begin{aligned}
molecular\ weight\ (COCl_2) \quad &= \quad at.\ wt.\ (C) + at.\ wt.\ (O) \\
&\quad + 2 \times at.\ wt.\ (Cl) \\
&= \quad 12.0\ g/mole + 16.0\ g/mole \\
&\quad + 2 \times 35.5\ g/mole \\
&= \quad 99.0\ g/mole.
\end{aligned}
$$

Hence, the number of moles of gas is

$$moles = \frac{mass}{molecular\ weight} = \frac{1.1 \times 10^{11}\ g}{99.0\ g/mole} = 1.1 \times 10^9\ moles.$$

Multiplying the number of moles by Avogadro's number, we obtain the number of molecules of gas:

$$
\begin{aligned}
\text{number of molecules} \;=\;& \text{moles} \times \text{Avogadro's number} \\
=\;& (1.1 \times 10^9 \text{ moles}) \times \\
& (6.02 \times 10^{23} \text{ molecules/mole}) \\
=\;& 6.6 \times 10^{32} \text{ molecules.}
\end{aligned}
$$

2.3 Molecular Weight and Formula Weight

The formula weight (molecular weight) of a molecule or compound is determined by the addition of its component atomic weights.

Example: FW of $CaCO_3$ = 1(40.1) + 1(12.0) + 3(16.0) = 100.1 g/mole.

Molecular weight = density × volume per molecule × Avogadro's number.

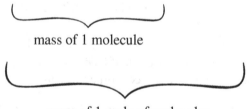

mass of 1 molecule

mass of 1 mole of molecules

2.4 Equivalent Weight

Equivalent weights are the amounts of substances that react completely with one another in chemical reactions. In electrolysis reactions, the equivalent weight is defined as that weight that either receives or donates 1 mole of electrons (6.02×10^{23} electrons) at an electrode.

For oxidation-reduction reactions, an equivalent is defined as the quantity of a substance that either gains or loses 1 mole of electrons (e^-).

In acid-base reactions, an equivalent of an acid is defined as the quantity of an acid that supplies 1 mole of H^+. An equivalent of a base supplies 1 mole of OH^-.

Note that a given substance may have any of several equivalent weights, depending on the particular reaction in which it is involved. For example, for Fe^{3+}:

$Fe^{3+} + e^- = Fe^{2+}$

one equivalent per mole;
eq. wt. = 55.8 g/eq.

$Fe^{3+} + 3e^- = Fe^0$

three equivalents per mole;

$$eq.\ wt. = \left(\frac{1}{3}\right)(55.8)$$

$$= 18.6\ g/eq.$$

Problem Solving Example:

Q In acting as a reducing agent, a piece of metal, M, weighing 16.0 g gives up 2.25×10^{23} electrons. What is the weight of one equivalent of the metal?

A One equivalent of a reducing agent is defined as that mass of the substance that releases the Avogadro number of electrons. Avogadro's number is 6.02×10^{23}; thus, one can find the number of equivalents in 16.0 g of the metal by dividing 2.25×10^{23} by 6.02×10^{23}.

$$number\ of\ eq. = \frac{2.25 \times 10^{23}\ electrons}{6.02 \times 10^{23}\ electrons/eq.} = 0.374\ eq.$$

Thus, there are 0.374 eq. weight per 16.0 g, and one can find the weight of one equivalent by dividing 16.0 g by 0.374 eq.

$$weight\ of\ 1\ eq. = \frac{16.0\ g}{0.374\ eq.} = 42.8\ g/eq.$$

2.5 Balancing Chemical Equations

When balancing chemical equations, one must make sure that there are the same number of atoms of each element on both the left and the right side of the arrow. For example, $H_2 + O_2 \rightarrow H_2O$ is not a balanced equation because there are two O's on the left side and only one on the right. $2H_2 + O_2 \rightarrow 2H_2O$ is the balanced equation for water because there are the same number of H and O atoms on each side of the equation.

Example:

$$2\,NaOH + H_2SO_4 \rightarrow Na_2SO_4 + 2H_2O$$

$$\left(\begin{array}{c} Na: 2\ atoms \\ O: 6\ atoms \\ H: 4\ atoms \\ S: 1\ atom \end{array}\right)$$

Problem Solving Example:

Q Balance the equations:

(a) $Ag_2O \rightarrow Ag + O_2$

(b) $Zn + HCl \rightarrow ZnCl_2 + H_2$

(c) $NaOH + H_2SO_4 \rightarrow Na_2SO_4 + H_2O$

A (a) $Ag_2O \rightarrow Ag + O_2$ is not a balanced equation because there are two Ag's on the left and only one on the right, and because there is only one O on the left and two O's on the right. To balance this equation, one must first multiply the left side by 2 to have two O's on each side.

$$2Ag_2O \rightarrow Ag + O_2$$

There are now four Ag's on the left and only one on the right. Therefore, the Ag on the right must be multiplied by 4.

$$2Ag_2O \rightarrow 4Ag + O_2$$

The equation is now balanced.

(b) $Zn + HCl \rightarrow ZnCl_2 + H_2$

In this equation, there are two H's and two Cl's on the right and only one of each on the left; therefore, the equation can be balanced by multiplying the HCl on the left by 2.

$$Zn + 2HCl \rightarrow ZnCl_2 + H_2$$

Because there are the same numbers of Zn, Cl, and H on both sides of the equation, it is balanced.

(c) $NaOH + H_2SO_4 \rightarrow Na_2SO_4 + H_2O$

Here, there are one Na, five O's, three H's, and one S on the left, and two Na's, one S, five O's, and two H's on the right. To balance this equation, one can first adjust the Na by multiplying the NaOH by 2.

$$2NaOH + H_2SO_4 \rightarrow Na_2SO_4 + H_2O$$

There are now two Na's, six O's, four H's, and one S on the left, and two Na's, five O's, two H's, and one S on the right. Because there are two more H's and one more O on the left than on the right, you can balance this equation by multiplying the H_2O by 2.

$$2NaOH + H_2SO_4 \rightarrow Na_2SO_4 + 2H_2O$$

The equation is now balanced.

2.6 Calculations Based on Chemical Equations

The coefficients in a chemical equation provide the ratio in which moles of one substance react with moles of another.

Example:

$$C_2H_4 + 3O_2 \rightarrow 2CO_2 + 2H_2O \text{ represents}$$

1 mole C_2H_4 + 3 moles $O_2 \rightarrow$ 2 moles CO_2 + 2 moles H_2O.

In this equation, the number of moles of O_2 consumed is always equal to three times the number of moles of C_2H_4 that react.

Problem Solving Example:

Q A metal has an atomic weight of 24 g/mole. When it reacts with a nonmetal of atomic weight 80 g/mole, it does so in a ratio of 1 atom to 2 atoms, respectively. With this information, how many grams of nonmetal will combine with 33 g of metal? If 1.0 g of metal is reacted with 5.0 g of nonmetal, find the amount of product produced.

A To answer this problem, write out the reaction between the metal and nonmetal, so that the relative number of moles that react can be determined. You can calculate the number of grams of material that react or are produced. You are told that 1 atom of metal reacts with 2 atoms of nonmetal. Let X = nonmetal and M = metal. The compound is MX_2. The reaction is $M + 2X \rightarrow MX_2$. Determine the number of moles that react. You have 33 g of M with an atomic weight of 24 g/mole.

Therefore, the number of moles $= \dfrac{33 \text{ g}}{24 \text{ g}/\text{mole}} = 1.375$ moles.

The above reaction states that for every 1 mole of M, 2 moles of X must be present. This means, therefore, that $2 \times 1.375 = 2.75$ moles of X must be present. The nonmetal has an atomic weight of 80 g/mole. Thus, recalling the definition of a mole, $2.75 \text{ mole} = \dfrac{\text{g}}{80 \text{ g/mole}}$. Solving for grams of X, you obtain $2.75(80) = 220$ g.

Let us consider the reaction of 1.0 g of M with 5.0 g of X to produce an unknown amount of MX_2. The solution is similar to the other, except that here you consider the concept of a limiting reagent. The amount of MX_2 produced from a combination will depend on the substance that exists in the smallest quantity. Thus, to solve this problem you compute the number of moles of M and X present. The smaller number (based on reaction equation) is the one you employ in calculating the number of moles of MX_2 that will be generated. You have, therefore,

$$M_{\text{moles}} = \frac{1.0}{24} = 0.04166 \text{ moles } M$$

$$X_{\text{moles}} = \frac{5.0}{80} = 0.0625 \text{ moles } X$$

Using 0.0625 moles of X, only 0.03125 moles of MX_2 will be produced, since the equation informs you that 1 mole of MX_2 is produced for every 2 moles of X. For M, it is a 1:1 ratio, so that 0.04166 moles of MX_2 would be generated. Therefore, X is the limiting reagent. The atomic weight of MX_2 is 184 g/mole. Thus, the amount produced is

$$184 \text{ g/mole } (0.03125 \text{ moles}) = 5.8 \text{ g.}$$

2.7 Limiting-Reactant Calculations

The reactant that is used up first in a chemical reaction is called the limiting reactant, and the amount of product is determined (or limited) by the limiting reactant.

Problem Solving Example:

 What is the maximum weight of SO_3 that could be made from 25.0 g of SO_2 and 6.0 g of O_2 by the following reaction?

$$2SO_2 + O_2 \rightarrow 2SO_3$$

 From the reaction, one knows that for every 2 moles of SO_3 formed, 2 moles of SO_2 and 1 mole of O_2 must react. Thus, to find the amount of SO_3 that can be formed, one must first know the number of moles of SO_2 and O_2 present. The number of moles is found by dividing the number of grams present by the molecular weight (MW).

$$\text{number of moles} = \frac{\text{number of grams}}{\text{MW}}$$

For O_2: MW = 32.0 g/mole.

$$\text{number of moles} = \frac{6.0\,g}{32.0\,g/mole} = 1.88 \times 10^{-1} \text{ moles}$$

For SO_2:   MW = 64.1 g/mole.

$$\text{number of moles} = \frac{25.0\,g}{64.1\,g/mole} = 3.90 \times 10^{-1} \text{ moles}$$

Because 2 moles of SO_2 are needed to react with 1 mole of O_2, 3.76×10^{-1} moles of SO_2 will react with 1.88×10^{-1} moles of O_2. This means that $3.90 \times 10^{-1} - 3.76 \times 10^{-1}$ moles, or 0.14×10^{-1} moles, of SO_2 will remain unreacted. In this case, O_2 is called the limiting reactant because it determines the number of moles of SO_3 formed. There will be twice as many moles of SO_3 formed as there are O_2 reacting.

$$
\begin{aligned}
\text{number of moles of } SO_3 \text{ formed} \; &= \; 2 \times 1.88 \times 10^{-1} \text{ moles} \\
&= \; 3.76 \times 10^{-1} \text{ moles}
\end{aligned}
$$

The weight is found by multiplying the number of moles formed by the molecular weight (MW of SO_3 = 80.1 g/mole).

weight of SO_3 = 3.76 × 10⁻¹ moles × 80.1 g/mole = 30.1 g

2.8 Theoretical Yield and Percentage Yield

The theoretical yield of a given product is the maximum yield that can be obtained from a given reaction if the reaction goes to completion (rather than to equilibrium).

The percentage yield is a measure of the efficiency of the reaction. It is defined as

$$\text{percentage yield} = \frac{\text{actual yield}}{\text{theoretical yield}} \times 100\%.$$

Problem Solving Example:

 Fifteen grams of ethane (MW = 30.0 g/mole) react with chlorine to yield 15 g of ethyl chloride (MW = 64.5 g/mole). What is the percent yield of ethyl chloride?

 The reaction is

$$Cl_2 + C_2H_6 \rightarrow C_2H_5Cl + HCl.$$

1 mole of ethane is (2 × 12.0) + (6 × 1.0) = 30.0 g.

From the equation, 1 mole of C_2H_6 yields 1 mole of C_2H_5Cl.

1 mole of C_2H_5Cl weighs (2 × 12.0) + (5 × 1.0) + (1 × 35.5) = 64.5 g.

Thirty grams of ethane theoretically should give 64.5 g of ethyl chloride. The amount of ethyl chloride obtained from 15 g of ethane is

$$\text{amount of } C_2H_5Cl = \frac{15}{30.0} \times 64.5 \text{ g} = 32 \text{ g}.$$

The theoretical yield is 32 g, but the actual yield is 15 g.

$$\text{percent yield} = \frac{15}{32} \times 100 = 47\%$$

2.9   Percentage Composition

The percentage composition of a compound is the percentage of the total mass contributed by each element:

$$\% \text{ composition} = \frac{\text{mass of element in compound}}{\text{mass of compound}} \times 100\%.$$

Problem Solving Example:

 Using the Periodic Table of Elements, find the following for sodium dihydrogen phosphate, NaH_2PO_4: (a) formula weight, (b) percent composition of oxygen, (c) weight in grams of 2.7 moles, and (d) percentage composition of oxygen in 2.7 moles.

A This problem encompasses work in chemical stoichiometry. With this in mind, you proceed as follows:

(a) The formula weight = molecular weight, which is the sum of atomic masses of all the atoms in the substance. Na = 22.99, H = 1.008, O = 15.999, and P = 30.974. Thus, formula weight = Na + 2H + P + 4O = 22.99 + 2(1.008) + 30.974 + 4(15.999) = 120.0.

(b) Percentage composition of oxygen is

$$\frac{\text{total mass of oxygen in compound}}{\text{total mass of compound}}$$

$$= \frac{4 \text{ oxygen atoms} \times 15.999 \text{ mass/oxygen atom}}{120.0}$$

$$= 0.5333 \text{ g of oxygen of } 53.33\% \text{ by mass in } NaH_2PO_4.$$

(c) $Mole = \frac{\text{mass of substance}}{\text{molecular weight}}$. You are given that there are 2.7 moles and you calculated the molecular weight. Thus, the mass in grams of 2.7 moles $= (120.0 \text{ g/mole})(2.7 \text{ moles}) = 324 \text{ g of } NaH_2PO_4$.

(d) Following a procedure similar to the one in part (b),

$$\text{percent composition} = \frac{\text{no. of moles of O} \times \text{MW of O}}{\text{mass of comp/no. of moles}} \times 100$$

$$= \frac{4 \text{ moles} \times 15.90 \text{ g/mole}}{324 \text{ g/2.7 mole}} \times 100$$

$$= 53.33\%.$$

The percent composition of any element in any compound does not change when the amount of the compound present is changed.

2.10 Density and Molecular Weight

At STP, standard temperature and pressure (0 °C and 760 mm of mercury pressure), 1 mole of any ideal gas occupies 22.4 L. (The "molar volume" of the gas at STP is 22.4 L/mole.)

The density can be converted to molecular weight using the 22.4 L/mole relationship:

$$MW = (\text{density})(\text{molar volume})$$

$$\left(\frac{g}{mole}\right) = \left(\frac{g}{L}\right)\left(\frac{L}{mole}\right)$$

Problem Solving Example:

Q If the density of ethylene is 1.25 g/L at STP and it is composed entirely of carbon and hydrogen in a ratio of 1:2, what is the molecular weight and formula of ethylene?

A This problem is solved once you know that at STP 1 mole of any gas occupies 22.4 L. Assume that 1 mole of ethylene gas is present. Density = mass/volume. As such the mass of the gas = (1.25 g/L)(22.4 L) = 28.0 g. Therefore, 1 mole of ethylene weighs 28 g.

From the ratio given in the question, you know the molecular formula can be represented as $(CH_2)_x$. To obtain the actual molecular formula, look for a compound that has a molecular weight of 28 g yet maintains the carbon:hydrogen ratio of 1:2. By looking at the Periodic Table for atomic weights and through some arithmetic, you will find that the only formula that meets these requirements is $(CH_2)_2$, or C_2H_4. This formula can also be found by dividing the weight of one CH_2 into 28 g. This solves for x in the expression $(CH_2)_x$. MW of CH_2 = 14 g/mole.

$$\text{number of } CH_2 = \frac{28 \text{ g}}{14 \text{ g/}CH_2} = 2CH_2 = C_2H_2$$

2.11 Weight-Volume Relationships

For a typical weight-volume problem, follow these steps:

Step 1: Write the balanced equation for the reaction.

Step 2: Write the given quantities and the unknown quantities for the appropriate substances.

Step 3: Calculate reacting weights or number of moles (or volume, if the reaction involves only gases) for the substances whose quantities are given. Make sure that the units for each substance are identical.

Step 4:   Use the proportion method or the factor-label method.

Step 5:   Solve for the unknown.

Example:

NaClO$_3$, when heated, decomposes to NaCl and O$_2$. What volume of O$_2$ at STP results from the decomposition of 42.6 g NaClO$_3$?

1. Balanced equation (using reactive masses):

$$213 \text{ g NaClO}_3 \xrightarrow{\Delta} 117 \text{ g NaCl} + 96 \text{ g O}_2$$

$$\frac{\text{mass O}_2 \text{ produced}}{\text{mass NaClO}_3 \text{ decomposed}} = \frac{96 \text{ g}}{213 \text{ g}} = \frac{x}{42.6 \text{ g}}$$

$$x = \frac{(96 \text{ g})(42.6 \text{ g})}{213 \text{ g}} = 19.2 \text{ g O}_2$$

2. Balanced equation (using moles):

$$2\text{NaClO}_3 \xrightarrow{\Delta} 2\text{NaCl} + 3\text{O}_2$$

$$\frac{\text{mass O}_2 \text{ produced}}{\text{mass NaClO}_3 \text{ decomposed}} = \frac{3}{2} = \frac{y}{(42.6 \text{ g} / 106.5 \text{ g} / \text{mole})}$$

$$y = \frac{(3)(42.6)}{(2)(106.5)} = 0.6 \text{ mole O}_2.$$

Finally, at STP, 1 mole (32.0 g) O$_2$ occupies 22.4 L, so from 1,

$$V_{O_{2(STP)}} = \left(\frac{22.4 \text{ L/mole}}{32.0 \text{ g/mole}} \right)(19.2 \text{ g}) = 13.4 \text{ L}$$

or from 2,

$$V_{O_{2(STP)}} = (0.6 \text{ mole})(22.4 \text{ L/mole}) = 13.4 \text{ L}.$$

The ideal gas law, $PV = nRT$, can be used to determine the volume of gas or the number of moles of a gas at conditions other than STP.

Problem Solving Examples:

 What volume of hydrogen at STP is produced when sulfuric acid reacts with 120 g of metallic calcium? The equation for the reaction is

$$Ca + H_2SO_4 \rightarrow CaSO_4 + H_2.$$

 This problem may be solved by using either the mole method or the proportion method.

Factor-label method: In using the mole method, one looks at this equation and sees that for every mole of calcium acted on by H_2SO_4, 1 mole of hydrogen is produced.

This means that to find how much hydrogen is produced, one must first find out how much calcium is present. There will be the same number of moles of hydrogen produced as there are moles of calcium reacted. After one knows how many moles of hydrogen are produced, one can calculate the volume.

To calculate the number of moles of calcium present, one must divide the amount present by the molecular weight (molecular weight of Ca = 40.1 g/mole).

$$\text{number of moles} = \frac{\text{number of grams present}}{\text{molecular weight}}$$

$$\text{number of moles of Ca} = \frac{120 \text{ g}}{40.1 \text{ g/mole}} = 3.0 \text{ moles}$$

Therefore, 3.0 moles of hydrogen gas are produced. At STP, the volume of 1 mole of any gas occupies 22.4 L. Thus, when 3.0 moles of gas are generated, as in this problem, it occupies $3.0 \times 22.4 \text{ L} = 67 \text{ L}$.

Proportion method: In the proportion method, the molecular weights (multiplied by the proper coefficients) are placed below the formula in the equation and the amounts of substances (given and unknown) are placed above. In this case, because one is trying to find the volume of hydrogen and not its weight, the volume of 1 mole will be placed below the equation as shown.

120 g X (Vol.)

$$Ca + H_2SO_4 \rightarrow CaSO_4 + H_2$$

40.1 g   22.4 L

$$\frac{120\ g}{40.1\ g/mole} = \frac{x}{22.4\ L}$$

X = unknown volume of H_2 produced.

Solving for X:

$$X = \frac{(22.4\ L)(120\ g)}{40.1\ g} = 67\ L$$

Quiz: Stoichiometry, Chemical Arithmetic

1. What is the gram-formula weight of $Na_2Cr_2O_7$?

    (A) 174 g   (D) 262 g

    (B) 196 g   (E) 286 g

    (C) 240 g

2. The number of atoms in a molecule of $Pb(Cr_2O_7)_2$ is

    (A) 3.   (D) 17.

    (B) 10.   (E) 19.

    (C) 14.

3. Consider the following unbalanced equation. Coefficients are missing.

    ____NH_3 + ____O_2 → ____NO + ____H_2O

    To balance the equation, the four consecutive coefficients from left to right are

    (A) 4, 5, 4, 6.   (D) 5, 5, 4, 6.

    (B) 4, 4, 5, 6.   (E) 6, 5, 4, 4.

    (C) 5, 4, 5, 6.

4. How many grams of oxygen are needed for the complete combustion of 39.0 g of C_6H_6? The molecular weight of C_6H_6 is 78.0 g/mole.

$$2C_6H_6 + 15O_2 \rightarrow 12CO_2 + 6H_2O$$

   (A) 3.75 g   (D) 60.0 g

   (B) 120.0 g   (E) 292.5 g

   (C) 32.0 g

5. What is the maximum weight of SO_3 that could be made from 40.0 g of SO_2 and 8.0 g of O_2 by the reaction below?

$$2SO_2 + O_2 \rightarrow 2SO_3$$

   (A) 40.0 g   (D) 20.0 g

   (B) 31.3 g   (E) 52.5 g

   (C) 25.0 g

6. If 2.0 grams of $NiCl_2$ is mixed in aqueous solution with 3.0 g NaOH to yield 0.019 mole of $Ni(OH)_2$, what is the percentage yield?

   (A) 43%   (D) 50%

   (B) 90%   (E) None of the above.

   (C) 85%

7. A sample containing aluminum weighing 10.0 grams yielded 2.0 grams of aluminum sulfide. What is the percentage of aluminum (atomic weight 27.0) in the sample?

   (A) $\dfrac{2.0}{10.0} \times 100$

   (B) $\dfrac{2.0}{10.0} \times \dfrac{27.0}{150.3} \times 100$

   (C) $\dfrac{2.0}{10.0} \times \dfrac{150.3}{3 \times 27.0} \times 100$

   (D) $\dfrac{2.0}{10.0} \times \dfrac{2 \times 27.0}{150.3} \times 100$

   (E) None of the above.

8. How many grams of Na are present in 30 g of NaOH?

   (A) 10 g   (D) 20 g

   (B) 15 g   (E) 22 g

   (C) 17 g

9. What is the density of a diatomic gas, measured at STP conditions, whose gram-molecular weight is 80 g?

   (A) 1.9 g/L   (D) 4.3 g/L

   (B) 2.8 g/L   (E) 5.0 g/L

   (C) 3.6 g/L

10. In the equation $Cu^{2+} + 2e^- = Cu$, the equivalent weight of Cu^{2+} is

    (A) 2.0 g/eq.   (D) 63.5 g/eq.

    (B) 21.2 g/eq.   (E) 127.0 g/eq.

    (C) 31.8 g/eq.

ANSWER KEY

1.	(D)	6.	(E)
2.	(E)	7.	(D)
3.	(A)	8.	(C)
4.	(B)	9.	(C)
5.	(A)	10.	(C)

CHAPTER 3

Atomic Structure and the Periodic Table

3.1   Atomic Spectra

The ground state is the lowest energy state available to the atom.

The excited state is any state of energy higher than that of the ground state.

The formula for changes in energy (ΔE) is

$$\Delta E_{\text{electron}} = E_{\text{final}} - E_{\text{initial}}.$$

When the electron moves from the ground state to an excited state, it absorbs energy.

When it moves from an excited state to the ground state, it emits energy.

This exchange of energy is the basis for atomic spectra.

3.2 The Bohr Theory of the Hydrogen Atom

Bohr applied to the hydrogen atom the concept that the electron can exist in only certain stable energy levels and that when the electronic state of the atom changes, it must absorb or emit exactly that amount of energy equal to the difference between the final and initial states: $\Delta E = E_a - E_b$.

$$E_b - E_a = \frac{Z^2 e^2}{2a_0}\left[\frac{1}{n_a{}^2} - \frac{1}{n_b{}^2}\right]$$

measures the energy difference between states a and b where n = the (quantum) energy level, E = energy, e = charge on electron, a_0 = Bohr radius, and Z = atomic number.

Problem Solving Examples:

Q The accompanying figure shows an energy-level diagram. Make a comparison of the energy of the $n = 5$ to $n = 4$ transition for an electron with an atomic number of $Z = 3$, and the energy of the $n = 2$ to $n = 1$ transition for an electron with an atomic number of $Z = 2$.

A When an atom is in the ground state, it is at its lowest energy level, $n = 1$. If energy is added, the electrons become "excited" and move to a higher energy level; $n = 2, 3, 4 \ldots$. After a time, the electrons fall back to lower energy states and release the energy needed to move them to the higher energy levels. The equation

$$E_b - E_a = Z^2 \frac{e^2}{2a_0}\left(\frac{1}{n_a{}^2} - \frac{1}{n_b{}^2}\right)$$

where n = the (quantum) energy level, E = energy, e = charge on electron, a_0 = Bohr radius, and Z = atomic number, measures this energy release between energy levels. You are asked to compare the energy releases between two different atoms and energy levels. You do not need to know the actual energy released. Because $e^2/2a_0$ is a constant, it need not be evaluated. (It will cancel out when you compare.)

Therefore, for $n = 5$ to $n = 4$ you have

$$E_b - E_a = \frac{e^2}{2a_0}\left(\frac{1}{n_a^2} - \frac{1}{n_b^2}\right)(Z^2) = \frac{e^2}{2a_0}\left(\frac{1}{16} - \frac{1}{25}\right) \times 9.$$

$n = \infty$	$E_\infty = 0$
$n = 5$	$E_5 = -\frac{1}{25}(Z^2 e^2 / 2a_0)$
$n = 4$	$E_4 = -\frac{1}{16}(Z^2 e^2 / 2a_0)$
$n = 3$	$E_3 = -\frac{1}{9}(Z^2 e^2 / 2a_0)$
$n = 2$	$E_2 = -\frac{1}{4}(Z^2 e^2 / 2a_0)$
$n = 1$	$E_1 = -(Z^2 e^2 / 2a_0)$

Energy →

Energy-level diagram showing relative energies for an atom consisting of a nucleus with charge Z^+ plus one electron with various values of n.

For $n = 2$ to $n = 1$ you have

$$E_b - E_a = \frac{e^2}{2a_0}\left(\frac{1}{n_a^2} - \frac{1}{n_b^2}\right)(Z^2) = \frac{e^2}{2a_0}\left(\frac{1}{1} - \frac{1}{4}\right) \times 4.$$

By comparison, in the form of a ratio, one has

$$\frac{\dfrac{e^2}{2a_0}\left(\dfrac{1}{16} - \dfrac{1}{25}\right) \times 9}{\dfrac{e^2}{2a_0}\left(\dfrac{1}{1} - \dfrac{1}{4}\right) \times 4} = \frac{\left(\dfrac{1}{16} - \dfrac{1}{25}\right) \times 9}{\left(\dfrac{1}{1} - \dfrac{1}{4}\right) \times 4} = \frac{0.2025}{3}.$$

$\dfrac{0.2025}{3} \times 100\% = 6.8\%$. Thus, the energy involved in the $n = 5$ to $n = 4$ transition is 6.8% as large as the energy emitted in the $n = 2$ to $n = 1$ transition.

The Rydberg-Ritz equation permits calculation of the spectral lines of hydrogen:

$$\frac{1}{\lambda} = R\left(\frac{1}{n_a^2} - \frac{1}{n_b^2}\right)$$

where $R = 109{,}678$ cm^{-1} (Rydberg constant), n_a and n_b are the quantum numbers for states a and b, and λ is the wavelength of light emitted or absorbed.

Q The Rydberg-Ritz equation governing the spectral lines of hydrogen is $\dfrac{1}{\lambda} = R\left(\dfrac{1}{n_a^2} - \dfrac{1}{n_b^2}\right)$, where R is the Rydberg constant, n_a indexes the series under consideration ($n_a = 1$ for the Lyman series, $n_a = 2$ for the Balmer series, $n_a = 3$ for the Paschen series), $n_b = n_a + 1, n_a + 2, n_a + 3, \ldots$ indexes the successive lines in a series, and λ is the wavelength of the line corresponding to index n_b. Thus, for the Lyman series, $n_a = 1$ and the first two lines are 1,215.56 Å ($n_b = n_a + 1 = 2$) and 1,025.83 Å ($n_b = n_a + 2 = 3$). Using these two lines, calculate two separate values of the Rydberg constant. The actual value of this constant is $R = 109,678$ cm^{-1}.

A The first thing to do is to convert the wavelengths from Å to more manageable units, i.e., centimeters. Using the relationship 1 Å = 10^{-8} cm, the first two Lyman lines are 1,215.56 Å = 1,215.56 $\times 10^{-8}$ cm for $n_b = 2$, and 1,025.83 Å = 1,025.83 $\times 10^{-8}$ cm for $n_b = 3$. Solving the Rydberg-Ritz equation for R, one obtains

$$R = \left(\lambda \left(\frac{1}{n_a^2} - \frac{1}{n_b^2} \right) \right)^{-1}.$$

For the first line,

$$R = \left(\lambda \left(\frac{1}{n_a^2} - \frac{1}{n_b^2} \right) \right)^{-1} = \left(1{,}215.56 \times 10^{-8}\,\text{cm} \left(\frac{1}{1^2} - \frac{1}{2^2} \right) \right)^{-1}$$

$$= 109{,}689 \text{ cm}^{-1}.$$

For the second line,

$$R = \left(\lambda \left(\frac{1}{n_a^2} - \frac{1}{n_b^2} \right) \right)^{-1} = \left(1{,}025.83 \times 10^{-8}\,\text{cm} \left(\frac{1}{1^2} - \frac{1}{3^2} \right) \right)^{-1}$$

$$= 109{,}667 \text{ cm}^{-1}.$$

The first of these is 0.0100% greater than the true value, and the second is 0.0100% less than the true value.

Light behaves as if it were composed of tiny packets, or quanta, of energy (now called "photons").

$$E_{\text{photon}} = h\mu$$

where h is Planck's constant and μ is the frequency of light.

$$E = \frac{hc}{\lambda}$$

where c is the speed of light and λ is the wavelength of light.

Problem Solving Example:

How much energy is emitted by Avogadro's number of atoms, if they each emit a light wave of 400 nm wavelength?

Energy is related to wavelength in the following equation, $E = hc/\lambda$, where E is energy, h is Planck's constant (6.626×10^{-34} Js), c is the speed of light (3.0×10^8 m/sec), and λ is the wavelength. Using this equation, one calculates the energy emitted by each atom. To find the total amount of energy produced by Avogadro's number of atoms, multiply the amount of energy one atom emits by 6.02×10^{23}.

Solving for E: 1 nm $= 10^{-9}$ m.

$$E = \frac{hc}{\lambda} = \frac{\left(6.626 \times 10^{-34}\,\text{Js}\right)\left(3.0 \times 10^8\,\text{m/s}\right)}{\left(400 \times 10^{-9}\,\text{m}\right)}$$

$$= 4.97 \times 10^{-19}\,\text{J/atom}$$

Thus, the total amount of energy emitted is

$$E = (4.97 \times 10^{-19}\,\text{J/atom})(6.02 \times 10^{23}\,\text{atoms/mole})$$
$$= 2.99 \times 10^5\,\text{J/mole}.$$

The electron is restricted to certain energy levels in the atom, specifically

$$E = -\frac{A}{n^2}$$

where A is 2.18×10^{-11} erg, and n is the quantum number.

3.3   Electric Nature of Atoms

3.3.1   Basic Electron Charges

Cathode rays are made up of very small negatively charged particles named electrons. The cathode is the negative electrode and the anode is the positive electrode.

The nucleus is made up of small positively charged particles called protons and of neutral particles called neutrons. The proton mass is approximately equal to the mass of the neutron and is 1,837 times the mass of the electron.

3.3.2   Components of Atomic Structure

The number of protons and neutrons in the nucleus is called the mass number, which corresponds to the isotopic atomic weight. The atomic number is the number of protons found in the nucleus.

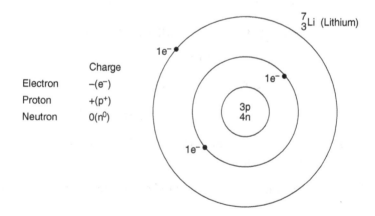

	Charge
Electron	$-(e^-)$
Proton	$+(p^+)$
Neutron	$0(n^0)$

^7_3Li (Lithium)

3p
4n

The electrons found in the outermost shell are called valence electrons. When these electrons are lost or partially lost (through sharing), the oxidation state is assigned a positive value for the element.

If valence electrons are gained or partially gained by an atom, its oxidation number is taken to be negative.

Problem Solving Example:

$\left(: \overset{\cdot\cdot}{\underset{\cdot\cdot}{Cl}} : \right)^{-}$ This is called the Lewis dot structure of the chloride ion. Its oxidation number is –1.

 Using Lewis-dot notation, show for each of the following the outer shell electrons for the uncombined atoms and for the molecules or ions that result:

(a)	H	+ H	→	hydrogen molecule
(b)	Br	+ Br	→	bromine molecule
(c)	Br	+ Cl	→	bromine chloride
(d)	Si	+ F	→	silicon fluoride
(e)	Se	+ H	→	hydrogen selenide
(f)	Ca	+ O	→	calcium oxide

When electrons are transferred from one atom to another, ions are formed, which gives rise to ionic bonding. Two atoms, both of which tend to gain electrons, may combine with each other by sharing one or more pairs of electrons. These two atoms form a covalent bond.

To solve this problem, one must know the number of valence electrons in each of the atoms in the equations. The valence number reflects the combining capacity of an atom. Next, one must know which atoms combine to form ionic bonds and which form covalent bonds. The only ionic bond formed in these equations is for Ca + O; the other bonds are covalent, and electrons are shared to form an isoelectronic electron cloud such as a noble gas.

Thus,

(a) H ·   +   H ·   →   H : H

(b) : Br :   +   : Br :   →   : Br : Br :

(c) : Br :   +   : Cl :   →   : Br : Cl :

(d) · Si ·   +   : F :   →   : F : Si : F :
(with : F : above and : F : below the central Si)

(e) H ·   +   · Se :   →   H : Se :
(with H below)

(f) · Ca ·   +   · O :   →   Ca⁺⁺ : O : ⁻⁻

3.4   The Wave Mechanical Model

Each wave function corresponds to a certain electronic energy and describes a region about the nucleus (called an orbital) where an electron having that energy may be found. The square of the wave function, $|\Psi|^2$, is called a probability density and equals the probability per unit volume of finding the electron in a given region of space.

Principal Quantum Number, n (Shell)	Azimuthal Quantum Number, ℓ (Subshell)	Subshell Designation	Magnetic Quantum Number, m (Orbital)	Number of Orbitals in Subshell
1	0	1s	0	1
2	0	2s	0	1
	1	2p	$-1,0,+1$	3
3	0	3s	0	1
	1	3p	$-1,0,+1$	3
	2	3d	$-2,-1,0,+1,+2$	5
4	0	4s	0	1
	1	4p	$-1,0,+1$	3
	2	4d	$-2,-1,0,+1,+2$	5
	3	4f	$-3,-2,-1,0,+1,+2,+3$	7

Table 3.1–Summary of Quantum Numbers

3.5   Subshells and Electron Configuration

The Pauli exclusion principle states that no two electrons within the same atom may have the same four quantum numbers.

main energy level	1	2	3	4
number of sublevels (n)	1	2	3	4
number of orbitals (n^2)	1	4	9	16
kind and no. of orbitals per sublevel	s 1	s p 1 3	s p d 1 3 5	s p d f 1 3 5 7
maximum no. of electrons per sublevel	2	2 6	2 6 10	2 6 10 14
maximum no. of electrons per main level ($2n^2$)	2	8	18	32

Table 3.2–Subdivision of Main Energy Levels

Problem Solving Example:

Q Write possible sets of quantum numbers for electrons in the second main energy level.

A In wave mechanical theory, four quantum numbers are needed to describe the electrons of an atom. The first, or principal, quantum number, n, designates the main energy level of the electron and has integral values of 1, 2, 3, ... The second quantum number, ℓ, designates the energy sublevel within the main energy level. The values of ℓ depend upon the value of n and range from zero to $n - 1$. The third quantum number, $m\ell$, designates the particular orbital within the energy sublevel. The number of orbitals of a given kind per energy sublevel is equal to the number of $m\ell$ values $(2\ell + 1)$. The quantum number $m\ell$ can have any integral value from $+\ell$ to $-\ell$ including zero. The fourth quantum number, s, describes the two ways in which an electron may be aligned with a magnetic field $\left(+\dfrac{1}{2} \text{ or } -\dfrac{1}{2} \right)$.

The states of the electrons within atoms are described by four quantum numbers, $n, \ell, m\ell$, and s. Another important factor is the Pauli exclusion principle which states that no two electrons within the same atom may have the same four quantum numbers.

To solve this problem, one must use the principles of assigning electrons to their orbitals.

If $n = 2$, ℓ can then have the values 0, 1; $m\ell$ can have the values of $+ 1, 0,$ or -1; and s is always $+\dfrac{1}{2}$ or $-\dfrac{1}{2}$. Thus, the answer is

$n = 2$

$\ell = 0, 1$

$m_\ell = 11, 0, -1$

$s = +\frac{1}{2}, -\frac{1}{2}$

	n	ℓ	m_ℓ	m_s
2s	2	0	0	$+\frac{1}{2}$
	2	0	0	$-\frac{1}{2}$
2p	2	1	$+1$	$+\frac{1}{2}$
	2	1	$+1$	$-\frac{1}{2}$
	2	1	0	$+\frac{1}{2}$
	2	1	0	$-\frac{1}{2}$
	2	1	-1	$+\frac{1}{2}$
	2	1	-1	$-\frac{1}{2}$

Main Levels	1	2			3	Summary
Sublevels	s	s	p		s	
H	↑					$1s^1$
He	↑↓					$1s^2$
Li	↑↓	↑				$1s^2 2s^1$
Be	↑↓	↑↓				$1s^2 2s^2$
B	↑↓	↑↓	↑ O O			$1s^2 2s^2 2p^1$
C	↑↓	↑↓	↑ ↑ O			$1s^2 2s^2 2p^2$
N	↑↓	↑↓	↑ ↑ ↑			$1s^2 2s^2 2p^3$
O	↑↓	↑↓	↑↓ ↑ ↑			$1s^2 2s^2 2p^4$
F	↑↓	↑↓	↑↓ ↑↓ ↑			$1s^2 2s^2 2p^5$
Ne	↑↓	↑↓	↑↓ ↑↓ ↑↓			$1s^2 2s^2 2p^6$
Na	↑↓	↑↓	↑↓ ↑↓ ↑↓		↑	$1s^2 2s^2 2p^6 3s^1$
Mg	↑↓	↑↓	↑↓ ↑↓ ↑↓		↑↓	$1s^2 2s^2 2p^6 3s^2$

Table 3.3–Electron Arrangements

The Hund rule states that for a set of equal-energy orbitals, each orbital is occupied by one electron before any orbital has two. Therefore, the first electrons to occupy orbitals within a sublevel have parallel spins. The rule is shown in Table 3.3.

 Apply the Hund rule to obtain the electron configuration for Si, P, S, Cl, and Ar.

Si	$3p^2$	↑	↑	
P	$3p^3$	↑	↑	↑
S	$3p^4$	↑↓	↑	↑
Cl	$3p^5$	↑↓	↑↓	↑
Ar	$3p^6$	↑↓	↑↓	↑↓

A The ground state of an atom is that in which the electrons are in the lowest possible energy level. Each level may contain two electrons of opposite spin. When there are several equivalent orbitals of the same energy, Hund rules are used to decide how the electrons are to be distributed between the orbitals:

(1) If the number of electrons is equal to or less than the number of equivalent orbitals, then the electrons are assigned to different orbitals.

(2) If two electrons occupy two different orbitals, their spins will be parallel in the ground state. The Hund rules states that the electrons attain positions as far apart as possible, which minimizes the repulsion obtained from interelectronic forces.

To solve this problem one must:

(1) find the total number of electrons within the atom,

(2) determine the number of valence electrons, and

(3) find the number of electrons in the highest equivalent energy orbital.

The total number of electrons in an atom is equal to that atom's atomic number.

Thus,

Si has 14 electrons

P has 15 electrons

S has 16 electrons

Cl has 17 electrons

Ar has 18 electrons

Next, from the orbital configuration:

$$_{14}Si \quad 1s^22s^22p^63s^2 \quad \cdot \quad 3p^2$$

$$_{15}P \quad 1s^22s^22p^63s^2 \quad \cdot \quad 3p^3$$

$$_{16}S \quad 1s^22s^22p^63s^2 \quad \cdot \quad 3p^4$$

$$_{17}Cl \quad 1s^22s^22p^63s^2 \quad \cdot \quad 3p^5$$

$$_{18}Ar \quad 1s^22s^22p^63s^2 \quad \cdot \quad 3p^6$$

One knows that the highest equivalent energy orbital is the 3p orbital. The number of electrons in this orbital increases by 1 starting with 2 for Si, then 3 for P, 4 for S, 5 for Cl, and 6 for Ar. Thus, Ar closes this orbital.

3.6 Isotopes

If atoms of the same element (i.e., having identical atomic numbers) have different masses, they are called isotopes.

The relative abundance of the isotopes is equal to their fraction in the element.

The average atomic weight, A, is equal to

$$A = X_1M_1 + X_2M_2 + \ldots + X_NM_N$$

where M_i is the atomic mass of isotope "i" and X_i is the corresponding probability of occurrence.

There are slight differences in chemical behavior of the isotopes of an element. Usually these differences, called isotope effects, influence the rate of reaction rather than the kind of reaction.

Problem Solving Example:

 Chromium exists in four isotopic forms. The atomic masses and percent occurrences of these isotopes are listed in the following table:

Isotopic mass (amu)	Percent occurrence
50	4.31%
52	83.76%
53	9.55%
54	2.38%

Calculate the average atomic weight of chromium.

A We will make use of the definition of average,

$$A = X_1M_1 + X_2M_2 + \ldots + X_NM_N,$$

where A is the average value, M_i is the atomic mass of isotope "i," and X_i is the corresponding probability of occurrence. For the four isotopes of chromium, we have:

$$M_1 = 50 \text{ amu} \quad X_1 = 4.31\% \quad = 0.0431$$
$$M_2 = 52 \text{ amu} \quad X_2 = 83.76\% \quad = 0.8376$$
$$M_3 = 53 \text{ amu} \quad X_3 = 9.55\% \quad = 0.0955$$
$$M_4 = 54 \text{ amu} \quad X_4 = 2.38\% \quad = 0.0238$$

Hence, the average atomic weight of chromium is

$A = X_1M_1 + X_2M_2 + X_3M_3 + X_4M_4 = (0.0431 \times 50 \text{ amu}) + (0.8376 \times 52 \text{ amu}) + (0.0955 \times 53 \text{ amu}) + (0.0238 \times 54 \text{ amu})$

$= 2.16 \text{ amu} + 43.56 \text{ amu} + 5.06 \text{ amu} + 1.29 \text{ amu}$

$= 52.07 \text{ amu}.$

3.7 Transition Elements and Variable Oxidation Numbers

Transition elements are elements whose electrons occupy the "d" sublevel.

Transition elements can exhibit various oxidation numbers. An example of this is manganese, with possible oxidation numbers of $+2$, $+3$, $+4$, $+6$, and $+7$.

Groups IB through VIIB and Group VIII of the Periodic Table constitute the transition elements.

3.8 Periodic Table

Periodic law states that chemical and physical properties of the elements are periodic functions of their atomic numbers.

Vertical columns are called groups, each containing a family of elements possessing similar chemical properties.

The horizontal rows in the periodic table are called periods.

The elements lying in two rows just below the main part of the table are called the inner transition elements.

In the first of these rows are elements 58 through 71, called the lanthanides, or rare earths.

The second row consists of elements 90 through 103, called the actinides.

Group IA elements are called the alkali metals.

Group IIA elements are called the alkaline earth metals.

Group VIIA elements are called the halogens.

Group O elements are called the noble gases.

The metals in the first two groups are the light metals, and those toward the center are the heavy metals. The elements found along the dark line in the chart are called metalloids. They have characteristics of both metals and nonmetals. Some examples of metalloids are boron and silicon.

3.9   Properties Related to the Periodic Table

The most active metals are found in the upper left corner. The most active nonmetals are found in the upper right corner.

Metallic properties include high electrical conductivity, luster, generally high melting points, ductility (ability to be drawn into wires), and malleability (ability to be hammered into thin sheets). Nonmetals are uniformly very poor conductors of electricity, do not possess the luster of metals, and form brittle solids. Metalloids have properties intermediate between those of metals and nonmetals.

3.9.1   Atomic Radii

The atomic radius generally decreases across a period from left to right. The atomic radius increases from top to bottom in a group.

Problem Solving Example:

Q Assume that 90% of the electron density is representative of the volume of the atom. The following atomic radii have been obtained for seven elements across the second period of the periodic table and for six elements down through the first family:

H						
0.37						
Li	Be	B	C	N	O	F
1.23	0.89	0.80	0.77	0.74	0.74	0.72
Na						
1.57						
K						
2.03						
Rb						
2.16						
Cs						
2.35						

Explain the observed trends of these atomic radii.

A Atomic radii increase going down a family but decrease going across a period. To explain this, one must consider the change in the elements in going across a period or down a family and see how it could affect atomic radii. In moving across a period, the atomic number increases, which means the nuclear positive charge increases. But the principal quantum number of the outside electrons remains constant. This means that the outer electrons, assuming all other factors are constant, will remain at the same distance from the nucleus.

Not all factors are constant, though. With increasing atomic number, the nuclear positive charge increases, and thus the force pulling the electrons towards the nucleus increases. Therefore, the atomic radii decreases as one moves across a period. Moving down a family presents a similar case of increasing atomic number. Here, however, the principal quantum number is also increasing, which means the outermost electrons are farther and farther away from the nucleus. The principal quantum number has a stronger effect than atomic number in determining atomic radii. Thus, the radii increase going down a family.

3.9.2 Electronegativity

The electronegativity of an element is a number that measures the relative strength with which the atoms of the element attract valence electrons in a chemical bond. This electronegativity number is based on an arbitrary scale from 0 to 4. Metals have electronegativities less than 2. Electronegativity increases from left to right in a period and decreases from top to bottom in a group.

3.9.3 Ionization Energy

Ionization energy is defined as the energy required to remove an electron from an isolated atom in its ground state. As we proceed down a group, a decrease in ionization energy occurs. Proceeding across a period from left to right, the ionization energy increases. As we proceed to the right, base-forming properties decrease and acid-forming properties increase.

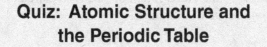

Quiz: Atomic Structure and the Periodic Table

1. The contribution of the electron to the atomic weight is

    (A) zero.

    (B) 1/1,837 that of a proton or a neutron.

    (C) equal to that of a proton.

    (D) equal to that of a neutron.

    (E) None of the above.

2. Element D has a valence of

    $$\cdot\overset{\displaystyle\cdot\cdot}{\underset{\displaystyle\cdot}{D}}\cdot$$

    (A) 1.   (D) 5.

    (B) 3.   (E) 7.

    (C) 4.

3. An atom has an atomic mass of 45 and an atomic number of 21. Select the correct statement about its atomic structure.

    (A) The number of electrons is 24.

    (B) The number of neutrons is 21.

    (C) The number of protons is 24.

    (D) The number of electrons and neutrons is equal.

    (E) The number of protons and neutrons is unequal.

4. According to the Hund rule, how many unpaired electrons does the ground state of iron have?

(A) 6   (D) 3

(B) 5   (E) 2

(C) 4

5. If p = protons, n = neutrons, and e = electrons, the relationship between these three atoms is that they are

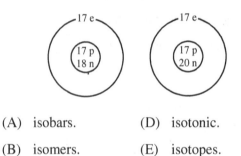

(A) isobars.   (D) isotonic.

(B) isomers.   (E) isotopes.

(C) isometric.

6. The element with atomic number 32 describes

(A) a metal.   (D) a halogen.

(B) a nonmetal.   (E) a noble gas.

(C) a metalloid.

7. Members of a common horizontal row of the periodic table should have the same

    (A) atomic number.

    (B) atomic mass.

    (C) electron number in the outer shell.

    (D) number of energy shells.

    (E) valence.

8. Which of the following are true statements with regard to the Periodic Table?

    (A) Electronegativity increases from left to right.

    (B) Ionization energy decreases from left to right.

    (C) Electronegativity increases from top to bottom.

    (D) Both (A) and (B).

    (E) (A), (B), and (C).

9. Which region of the Periodic Table represents the element with the largest atomic radius?

    (A) Upper left   (D) Lower right

    (B) Upper right   (E) Middle

    (C) Lower left

10. When an electron moves from the ground state to an excited state, it

 (A) absorbs energy.

 (B) emits energy.

 (C) has no change in energy.

 (D) Both (A) and (B).

 (E) None of the above.

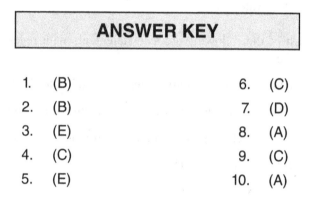

ANSWER KEY

1.	(B)		6.	(C)
2.	(B)		7.	(D)
3.	(E)		8.	(A)
4.	(C)		9.	(C)
5.	(E)		10.	(A)

CHAPTER 4

Bonding

4.1   Types of Bonds

An ionic bond occurs when one or more electrons are transferred from the valence shell of one atom to the valence shell of another.

The atom that loses electrons becomes a positive ion (cation), whereas the atom that acquires electrons becomes a negatively charged ion (anion). The ionic bond results from the electrostatic attraction between the oppositely charged ions.

The octet rule states that atoms tend to gain or lose electrons until there are eight electrons in their valence shell.

A covalent bond results from the sharing of a pair of electrons between atoms.

In a nonpolar covalent bond, the electrons are shared equally.

Nonpolar covalent bonds are characteristic of homonuclear diatomic molecules. For example, the fluorine molecule:

$$: \overset{\cdot\cdot}{\underset{\cdot\cdot}{F}} \cdot \quad + \quad \cdot \overset{\cdot\cdot}{\underset{\cdot\cdot}{F}} : \quad \rightarrow \quad : \overset{\cdot\cdot}{\underset{\cdot\cdot}{F}} : \overset{\cdot\cdot}{\underset{\cdot\cdot}{F}} :$$

Fluorine atoms → Fluorine molecule

When there is an unequal sharing of electrons between the atoms involved, the bond is called a polar covalent bond. An example:

$$
\begin{array}{l}
\quad\quad\; \overset{\cdot\cdot}{}\\
H \quad \overset{\times}{\underset{\cdot\cdot}{:}}Cl: \\
\end{array}
\quad
\begin{array}{l}
\times \text{ hydrogen electron}\\
\cdot \text{ chlorine electrons}
\end{array}
$$

$$
\begin{array}{l}
\quad\quad \overset{\cdot\cdot}{}\\
H \quad \overset{\times}{:}O: \\
\quad\quad \overset{\times\cdot}{}\\
\quad\quad H
\end{array}
\quad
\begin{array}{l}
\times \text{ hydrogen electron}\\
\cdot \text{ oxygen electrons}
\end{array}
$$

Because of the unequal sharing, the bonds shown are said to be polar bonds (dipoles). The more electronegative element in the bond is the negative end of the bond dipole. In each of the molecules shown here, there is also a non zero molecular dipole moment, given by the vector sum of the bond dipoles.

A pure crystal of elemental metal consists of roughly Avogadro's number of atoms held together by metallic bonds.

Problem Solving Example:

 Distinguish a metallic bond from an ionic bond and from a covalent bond.

 The best way to distinguish between these bonds is to define each and provide an illustrative example of each. When an actual transfer of electrons results in the formation of a bond, it can be said that an ionic bond is present. For example,

$$2K^0 \;+\; :\overset{\cdot\cdot}{\underset{\cdot}{S}}\cdot \;\rightarrow\; 2K^+ \;+\; :\overset{\cdot\cdot}{\underset{\cdot\cdot}{S}}:^{2-} \;\rightarrow\; K_2S$$

potassium atoms	sulfur atom	potassium ions (unlike ions due to transfer of electrons from potassium to sulfur)	sulfur ion	ionic bond due to the attraction of unlike ions

When a chemical bond is the result of the sharing of electrons, a covalent bond is present. For example:

$$: \overset{..}{\underset{..}{Br}} \cdot + \cdot \overset{..}{\underset{..}{F}} : \quad \rightarrow \quad : \overset{..}{\underset{..}{Br}} : \overset{..}{\underset{..}{F}} :$$

These electrons are shared by both atoms.

A pure crystal of elemental metal consists of millions of atoms held together by ionic bonds. Metals possess electrons that can easily ionize, i.e., they can be easily freed from the individual metal atoms. This free state of electrons in metals binds all the atoms together in a crystal. The free electrons extend over all the atoms in the crystal and the bonds formed between the electrons and positive nucleus are electrostatic in nature. The electrons can be pictured as a "cloud" that surrounds and engulfs the metal atoms.

4.2   Intermolecular Forces of Attraction

A dipole consists of a positive and negative charge separated by a distance. A dipole is described by its dipole moment, which is equal to the charge times the distance between the positive and negative charges:

net dipole moment = charge × distance.

In polar molecular substances, the positive pole of one molecule attracts the negative pole of another. The force of attraction between polar molecules is called a dipolar force.

When a hydrogen atom is bonded to a highly electronegative atom, it will become partially positively charged and will be attracted to neighboring electron pairs. This creates a hydrogen bond. The more polar the molecule, the more effective the hydrogen bond is in binding the molecules into a larger unit.

The relatively weak attractive forces between molecules are called van der Waals forces. These forces become apparent only when the molecules approach one another closely (usually at low temperatures and high pressure). They are due to the way the partially positive end of one molecule attract the partially negative end of another molecule. Compounds of the solid state, which are found mainly by this type of attraction, have soft crystals, are easily deformed, and vaporize easily. Because of the low intermolecular forces, the melting points are low and evaporation takes place so easily that it may occur at room temperature. Examples of substances with this last characteristic are iodine crystals and naphthalene crystals.

Problem Solving Example:

 Of the following pairs, which member should exhibit the largest dipole moment? Use the data from the accompanying table. (a) H–O and H–N; (b) H–F and H–Br; (c) C–O and C–S.

Pauling electronegativities
(H = 2.1)

1	2	3	4	5	6	7	8	9	10	11	12	13	14	15	16	17
Li 1.0	Be 1.5											B 2.0	C 2.5	N 3.0	O 3.5	F 4.0
Na 0.9	Mg 1.2											Al 1.5	Si 1.8	P 2.1	S 2.5	Cl 3.0
K 0.8	Ca 1.0	Sc 1.3	Ti 1.5	V 1.6	Cr 1.6	Mn 1.5	Fe 1.8	Co 1.8	Ni 1.8	Cu 1.9	Zn 1.6	Ga 1.6	Ge 1.8	As 2.0	Se 2.4	Br 2.8
Rb 0.8	Sr 1.0	Y 1.2	Zr 1.4	Nb 1.6	Mo 1.8	Tc 1.9	Ru 2.2	Rh 2.2	Pd 2.2	Ag 1.9	Cd 1.7	In 1.7	Sn 1.8	Sb 1.9	Te 2.1	I 2.5
Cs 0.7	Ba 0.9	La – Lu 1.1 – 1.2	Hf 1.3	Ta 1.5	W 1.7	Re 1.9	Os 2.2	Ir 2.2	Pt 2.2	Au 2.4	Hg 1.9	Tl 1.8	Pb 1.8	Bi 1.9	Po 2.0	At 2.2
Fr 0.7	Ra 0.9	Ac 1.1	Th 1.3	Pa 1.5	U 1.7	Np 1.3										

The values given in the table refer to the common oxidation states of the elements. For some elements, variation of the electronegativity with oxidation number is observed. For example, Fe(II) 1.8, Fe(III) 1.9; Cu(I) 1.9, Cu(II) 2.0; Sn(II) 1.8, Sn(IV) 1.9.

A A dipole consists of a positive and a negative charge separated by some distance. Quantitatively, a dipole is described by giving its dipole moment, which is equal to the charge times the distance between the positive and negative centers. The polarity of a bond is measured as the magnitude of the moment of the dipole. Thus, to find which member has the higher dipole moment, determine which bond has the greatest polarity. The polarity of the bond is indicated by the difference in electronegativities of the atoms (i.e., the difference in their tendency to attract shared electrons in a chemical covalent bond). The greater the difference in electronegativities, the greater the bond polarity, giving it the greater dipole moment. From the table, note that in part

(a) the difference in electronegativities of H–O is $3.5 - 2.1$, or 1.4. For H–N, the difference is $3.0 - 2.1$, or 0.9. Therefore, H–O has the largest dipole moment. Using similar calculations, one can determine the bonds with the larger dipole moments in parts (b) and (c).

(b) H−F and H−Br

For H−F: $4.0 - 2.1 = 1.9$

For H−Br: $2.8 - 2.1 = 0.7$

Thus, H−F has the larger dipole moment.

(c) C−O and C−S

For C−O: $3.5 - 2.5 = 1.0$

For C−S: $2.5 - 2.5 = 0$

Thus, C−O has the larger dipole moment.

4.3   Double and Triple Bonds

Sharing two pairs of electrons produces a double bond. An example:

$$\text{O} \overset{\times\times}{\underset{\times\times}{\overset{\times}{\times}}} : \text{C} : \overset{\times\times}{\underset{\times\times}{\overset{\times}{\times}}} \text{O} \quad \text{or} \quad \overset{..}{\text{O}} = \text{C} = \overset{..}{\text{O}}$$

The sharing of three electron pairs results in a triple bond. An example:

$$\text{H} \overset{\times}{.} \text{C} \overset{..}{::} \text{C} \overset{\times}{.} \text{H} \quad \text{or} \quad \text{H} - \text{C} \equiv \text{C} - \text{H}$$

Greater energy is required to break double bonds than single bonds, and triple bonds are harder to break than double bonds. Molecules that contain double and triple bonds have smaller interatomic distances and greater bond strength than molecules with only single bonds. Thus, in the series

$$H_3C - CH_3,\ H_2C = CH_2,\ HC \equiv CH,$$

the $C-C$ distance decreases, and the $C-C$ bond energy increases because of increased bonding.

4.4   Resonance Structures

The resonance structures for sulfur dioxide are as follows:

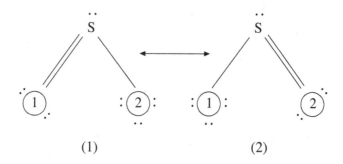

      (1)   (2)

The actual electronic structure of SO_2 does not correspond to either (1) or (2), but instead to an "average" structure somewhere in between. This true structure is known as a resonance hybrid of the contributing structures (1) and (2).

4.5   Electrostatic Repulsion and Hybridization

The process of mixing different orbitals of the same atom to form a new set of equivalent orbitals is termed hybridization. The orbitals formed are called hybrid orbitals.

Valence Shell Electron Pair Repulsion (VSEPR) theory permits the geometric arrangement of atoms, or groups of atoms, about some central atom to be determined solely by considering the repulsions between the electron pairs present in the valence shell of the central atom.

Based on VSEPR, the general shape of any molecule can be predicted from the number of bonding and nonbonding electron pairs in the valence shell of the central atom, recalling that nonbonded pairs of electrons (lone pairs) are more repellent than bonded pairs.

Number of Bonds	Number of Unused e Pairs	Type of Hybrid Orbital	Angle Between Bonded Atoms	Geometry	Example
2	0	sp	180°	Linear	BeF_2
3	0	sp^2	120°	Trigonal planar	BF_3
4	0	sp^3	109.5°	Tetrahedral	CH_4
3	1	sp^3	< 109.5°	Pyramidal	NH_3
2	2	sp^3	< 109.5°	Angular	H_2O
6	0	sp^3d^2	90°	Octahedral	SF_6

Table 4.1–Summary of Hybridization

Problem Solving Example:

 Describe the bonding in linear, covalent $BeCl_2$ and trigonal, planar, covalent BCl_3. What is the difference in the hybrid orbitals used?

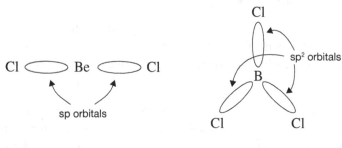

Figure A   Figure B

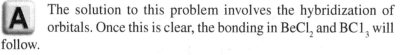 The solution to this problem involves the hybridization of orbitals. Once this is clear, the bonding in $BeCl_2$ and BCl_3 will follow.

Quantum theory deals with independent orbitals, such as 2s and 2p. This can be applied to a species, like hydrogen, with only one electron. However, with atoms that contain more than one electron, different methods must be used. For example, the presence of a 2s electron perturbs a 2p electron, and vice versa, such that a 2s electron makes a 2p electron take on some s-like characteristics. The result is that the hydrogen-like 2s and 2p orbitals are replaced by new orbitals, which contain the combined characteristics of the original orbitals. These new orbitals are called hybrid orbitals. The number of hybrid orbitals resulting from hybridization equals the number of orbitals being mixed together. For example, if one mixes an s and a p orbital, one obtains two sp hybrid orbitals. One s and two p = three sp^2 orbitals. One s and three p = four sp^3 orbitals. sp orbitals are linear. sp^2 orbitals assume a trigonal planar shape, and sp^3 orbitals assume a tetrahedral shape.

Solving: It is given that $BeCl_2$ is linear and covalent. Since sp orbitals are linear, Be undergoes sp hybridization. If something is linear, the bond angle is 180°. By understanding hybridization, one also knows the geometry of the molecule. A diagram of the bonding resembles Figure A.

Given that BCl_3 is trigonal planar, and since sp^2 hybridization yields a trigonal planar structure, B has sp^2 hybridized bonding with angles of 120° (Figure B).

4.6 Sigma and Pi Bonds

A molecular orbital that is symmetrical around the line passing through two nuclei is called a sigma (*s*) orbital. When the electron density in this orbital is concentrated in the bonding region between two nuclei, the bond is called a sigma (σ) bond.

The bond that is formed by the sideways overlap of two *p* orbitals, and that provides electron density above and below the line connecting the bound nuclei, is called a pi (π) bond.

Pi bonds are present in molecules containing double or triple bonds.

Of the sigma and pi bonds, the former has greater orbital overlap and is generally the stronger bond.

4.7 Properties of Ionic Substances

Ionic substances are characterized by the following properties:

1. Ionic crystals have large lattice energies because the electrostatic forces between them are strong.

2. In the solid phase, they are poor electrical conductors.

3. In the liquid phase, they are relatively good conductors of electric current; the mobile charges are the ions (in contrast to metallic conduction, where the electrons constitute the mobile charges).

4. They have relatively high melting and boiling points.

5. They are relatively nonvolatile and have low vapor pressure.

6. They are brittle.

7. Those that are soluble in water form electrolytic solutions that are good conductors of electricity.

4.8 Properties of Molecular Crystals and Liquids

The following are general properties of molecular crystals and/or liquids:

1. Molecular crystals tend to have small lattice energies and are easily deformed because their constituent molecules have relatively weak forces between them.

2. Both the solids and liquids are poor electrical conductors.

3. Many exist as gases at room temperature and atmospheric pressure; those that are solid or liquid at room temperature are relatively volatile.

4. Both the solids and liquids have low melting and boiling points.

5. The solids are generally soft and have a waxy consistency.

6. A large amount of energy is often required to chemically decompose the solids and liquids into simpler substances.

Quiz: Bonding

1. An ionic bond is best described as

    (A) an equal sharing of electrons.

    (B) an unequal sharing of electrons.

    (C) the gain of one or more electrons on one atom with the loss of one or more electrons on the other atom.

    (D) the attraction of one atom's nucleus to the electrons of another atom.

    (E) the mutual repulsion of a pair of electrons by two nuclei.

2. When the electrons of a bond are shared unequally by two atoms, the bond is said to be

    (A) covalent.   (D) ionic.

    (B) polar covalent.   (E) metallic.

    (C) coordinate covalent.

3. An example of a dipole molecule is

    (A) CH_4.   (D) NaCl.

    (B) H_2.   (E) O_2.

    (C) H_2O.

4. Which of the following exhibits hydrogen bonding?

    (A) NH_3 (D) Both (A) and (B).

    (B) CH_4 (E) (A), (B), and (C).

    (C) BH_3

5. Which of the following represents a resonance structure for CO_2?

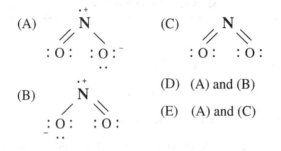

(A)

(C)

(B)

(D) (A) and (B)

(E) (A) and (C)

6. Which of the following molecular geometries is typical of sp bonding?

   (A) Tetrahedral

   (B) Trigonal planar

   (C) Linear

   (D) Trigonal bipyramidal

   (E) Octahedral

7. The VSEPR method is used to

   (A) predict the geometries of an atom.

   (B) estimate the energy levels of orbitals in an atom.

   (C) estimate electronegativities of elements.

   (D) predict the geometries of molecules and ions.

   (E) There is no such thing as a VSEPR method.

8. Which one of the following characteristics is not usually attributable to ionic substances?

   (A) High melting point

   (B) Brittleness

   (C) Crystalline (in the solid state)

   (D) Deforms when struck

   (E) Well-defined three-dimensional structure

9. A bond in which the electron density of a particular orbital is concentrated in the bonding region between two nuclei is known as a

 (A) single bond.   (D) sigma bond.

 (B) double bond.   (E) pi bond.

 (C) triple bond.

10. Which of the following is not a property of a molecular crystal?

 (A) Small lattice energy

 (B) Low melting point

 (C) Good electrical conductor

 (D) Easily deformed

 (E) Generally soft

ANSWER KEY

1.	(C)	6.	(C)
2.	(B)	7.	(D)
3.	(C)	8.	(D)
4.	(A)	9.	(D)
5.	(D)	10.	(C)

CHAPTER 5

Chemical Formulas

5.1 Chemical Formulas

A chemical formula is a representation of the makeup of a compound in terms of the kinds of atoms and their relative numbers.

5.2 Naming Compounds

Binary compounds consist of two elements. The name shows the two elements present and ends in -ide, such as $NaCl$ = sodium chloride. If the metal has only two possible oxidation states, use the suffix -ous for the lower one and -ic for the higher one.

Example: $FeCl_2$ = ferrous chloride [iron (II) chloride]

   $FeCl_3$ = ferric chloride [iron (III) chloride]

When naming binary covalent compounds formed between two nonmetals, a third system of nomenclature is preferred in which the numbers of each atom in a molecule is specified by a Greek prefix: di- (2), tri- (3), tetra- (4), penta- (5), and so on.

Example: N_2O_5 = dinitrogen pentoxide

Ternary compounds consist of three elements and are usually made up of an element and a polyatomic ion. To name these compounds you merely name each component, the positive one first and the negative one second.

Binary acids use the prefix hydro- in front of the stem or full name of the nonmetallic element and add the ending -ic.

Example: hydrochloric acid (HCl)

Ternary acids (oxyacids) usually contain hydrogen, a nonmetal, and oxygen. The most common form of the acid consists of only the stem of the nonmetal with the ending -ic. The acid containing one less atom of oxygen than the most common acid has the ending -ous. The acid containing one more atom of oxygen than the most common acid has the prefix per- and the ending -ic. The acid containing one less atom of oxygen than the -ous acid has the prefix hypo- and the ending -ous.

Salts produced by neutralization of acids contain polyatomic anions. The anion derived from the -ic acid ends in -ate, whereas the anion derived from the -ous acid ends in -ite.

Example: manganese (II) sulfate ($MnSO_4$)

Oxidation	Name of Acid		Examples	Name of Anion	
+1	hypo-	−ous	$HClO$, $HBrO$, HIO	hypo-	−ite
+3		−ous	$HClO_2$		−ite
+5		−ic	$HClO_3$, $HBrO_3$, HIO_3		−ate
+7	per-	−ic	$HClO_4$, $HBrO_4$, HIO_4, H_5IO_6	per-	−ate

Table 5.1–Group VIIA Oxyacids and Oxyions[a]

[a]The recently discovered HFO is not included because it has yet to be studied in detail.

5.2.1 Acid Salts

Partial neutralization of an acid that is capable of furnishing more than one H^+ per acid molecule produces salts that are called acid salts.

When only one acid salt is formed, the salt can be named by adding the prefix bi- to the name of the anion of the acid.

Example: $NaHSO_4$ = sodium bisulfate

The salt can also be named by specifying the presence of H, by writing hydrogen.

Example: Na_2HPO_4 = sodium hydrogen phosphate
(or disodium hydrogen phosphate)

5.3   Writing Formulas

5.3.1   General Observations

1. Metals are assigned positive oxidation numbers whereas non-metals (and all the polyatomic ions, except the ammonium ion) are assigned negative oxidation numbers.

2. A polyatomic ion is a group of elements that remain bonded as a group even when involved in formation of compounds.

5.3.2   Basic Rules for Writing Formulas

1. Represent the symbols of the components using the positive part first, and then the negative part:

     Sodium Chloride   Calcium Oxide   Ammonium Sulfate

     NaCl   CaO $(NH_4)_2SO_4$

2. Indicate the respective oxidation number, above and to the right of each symbol:

$$Na^{1+}\ Cl^{1-} \qquad\qquad Ca^{2+}\ O^{2-} \qquad\qquad (NH_4)^{1+}(SO_4)^{2-}$$

3. Write the subscript number equal to the oxidation number of the other element or radical:

$$Na^{1}Cl^{1-}, \quad Ca^{2+}O^{2-}, \quad (NH_4)^{1+}(SO_4)^{2-}$$

4. Now rewrite the formulas, omitting the subscript 1, the parentheses of the polyatomic ion which have the subscript 1, and the plus and minus signs:

$$NaCl \qquad\qquad Ca_2O_2 \qquad\qquad (NH_4)_2SO_4$$

5. As a general rule, the subscript numbers in the final formula are reduced to their lowest terms; hence, Ca_2O_2 becomes CaO. There are, however, certain exceptions, such as hydrogen peroxide (H_2O_2) and acetylene (C_2H_2).

5.4 Empirical and Molecular Formulas

The empirical formula of any compound gives the relative number of atoms of each element in the compound. It is the simplest formula of a material that can be derived solely from its components.

The molecular formula of a substance indicates the actual number of atoms in a molecule of the substance. To determine the molecular formula, you must calculate the empirical formula and then extrapolate to the molecular formula via the molecular weight. The molecular formula is a whole-number multiple of the empirical formula.

Problem Solving Example:

 When 10.00 g of phosphorus was reacted with oxygen, it produced 17.77 g of a phosphorus oxide. This phosphorus oxide was found to have a molecular weight of approximately 220 g/mole in the vapor state. Determine its molecular formula.

A The molecular formula of a substance indicates the relative number of atoms in a molecule of the substance. Therefore, to solve this problem you must first calculate the ratios of the moles to each other and the empirical formula, and then extrapolate the molecular formula via the molecular weight.

The number of moles of phosphorus (P) is

$$\frac{\text{wt. in grams of P}}{\text{atom weight}} = \frac{10.00}{31.0} = 0.3226 \text{ moles P.}$$

For oxygen, we have $\dfrac{7.777\text{g}}{16.0} = 0.486$ moles O.

The weight in grams of oxygen is 7.77 because the final product weighs 17.77 g and the phosphorus weighs 10.00 g. Since the only other element is oxygen, its weight must be the difference.

The ratio of the moles of P and O is 1:1.5, or 2:3. Therefore, the empirical formula of the oxide is P_2O_3.

To calculate the molecular formula, we must use the stated molecular weight of 220 g/mole. We must look for a formula that totals this molecular weight AND maintains the 2:3 ratio of P:O as expressed in the empirical formula. With some arithmetic, we find that the only formula that meets these two requirements is P_4O_6; 4:6 is the same as 2:3. The atomic weights of P and O are, respectively, 31.0 g/mole and 16.0 g/mole. We have four P atoms for a total of 124 g/mole, and we have six O atoms for a total of 96.0 g/mole. Now add $124 + 96 = 220$ g/mole.

Another method for determining the molecular formula is to divide the molecular weight of the molecule, 220 g/mole, by the weight of 1 P_2O_3, 110 g/mole.

$$\text{number of } P_2O_3 = \frac{220\text{g} / \text{mole}}{110\text{g} / \text{mole of } P_2O_3} = 2 \text{ moles of } P_2O_3$$

The formula is, therefore, $2 \times P_2O_3$, or P_4O_6.

CHAPTER 6

Types and Rates of Chemical Reactions

6.1 Types of Chemical Reactions

The four basic kinds of chemical reactions are: combination, decomposition, single replacement, and double replacement. ("Replacement" is sometimes called "metathesis.")

Combination can also be called synthesis. This refers to the formation of a compound from the union of its elements. For example:

$$Zn + S \rightarrow ZnS.$$

Decomposition, or analysis, refers to the breakdown of a compound into its individual elements and/or compounds. For example:

$$C_{12}H_{22}O_{11} \rightarrow 12\,C + 11H_2O.$$

The third type of reaction is called single replacement or single displacement. This type can best be shown by some examples where one substance is displacing another. For example:

$$Fe + CuSO_4 \rightarrow FeSO_4 + Cu.$$

The last type of reaction is called double replacement or double displacement, because there is an actual exchange of "partners" to form new compounds. For example:

$$AgNO_3 + NaCl \rightarrow AgCl + NaNO_3.$$

6.2   Measurements of Reaction Rates

The measurement of reaction rate is based on the rate of appearance of a product or disappearance of a reactant. It is usually expressed in terms of change in concentration of one of the participants per unit time:

$$\text{rate of chemical reation} = \frac{\text{change in concentration}}{\text{time}}$$

$$= \frac{\text{moles} / \text{L}}{\text{seconds}}.$$

For the general reaction $2AB \rightarrow A_2 + B_2$,

$$\text{average rate} = \frac{[AB]_2 - [AB]_1}{t_2 - t_1} = \frac{-\Delta[AB]}{\Delta t}$$

where $[AB]_2$ means "the AB concentration at time t_2 and $[AB]$ means "the initial concentration of AB."

Problem Solving Example:

For the following reaction A + B → C, at time = 0, $[C]$ = 0.25 mole/L. Ten seconds later $[C]$ = 0.75 mole/L. What is the rate of the reaction?

The rate of a chemical reaction is determined by the change in concentration (either the appearance of a product or disappearance of a reactant) per unit time such that:

$$\text{rate} = \frac{\Delta[C]}{\Delta t}$$

$$\text{rate} = \frac{0.75\,\text{moles}\,/\,\text{L} - 0.25\,\text{moles}\,/\,\text{L}}{10\,\text{seconds}} = \frac{0.05\,\text{moles}\,/\,\text{L}}{\text{seconds}}.$$

6.3   Factors Affecting Reaction Rates

There are five important factors that control the rate of a chemical reaction. These are summarized below.

1. The nature of the reactants and products, i.e., the nature of the transition state formed. Some elements and compounds, because of the bonds broken or formed, react more rapidly with each other than do others.

2. The surface area exposed. Since most reactions depend on the reactants coming into contact, increasing the surface area exposed proportionally increases the rate of the reaction.

3. The concentrations. The reaction rate usually increases with increasing concentrations of the reactants.

4. The temperature. A temperature increase of 10 °C above room temperature usually causes the reaction rate to double.

5. The catalyst. Catalysts speed up the rate of a reaction but do not change the equilibrium constant (i.e., they simply speed up the rate of approach to equilibrium).

6.4   The Arrhenius Equation: Relating Temperature and Reaction Rate

The following is the Arrhenius equation:

$$k = Ae^{-E_a/RT}$$

where k is the rate constant, A = the Arrhenius constant, E_a = activation energy, R = universal gas constant, and T = temperature in Kelvin. k is small when the activation energy is very large or when the temperature of the reaction mixture is low.

$$\ln k = \ln A - \frac{E_a}{RT}$$

$$\updownarrow \qquad \updownarrow \qquad \updownarrow\updownarrow$$

$$y = b + mx$$

A plot of $\ln k$ versus $1/T$ gives a straight line whose slope is equal to $-E_a/R$ and whose intercept with the ordinate is $\ln A$.

Problem Solving Example:

What activation energy should a reaction have so that raising the temperature by 10 °C at 0 °C would triple the reaction rate?

The activation energy is related to the temperature by the Arrhenius equation, which is stated

$$k = Ae^{-E_a/RT}$$

where A is a constant characteristic of the reaction, e is the base of natural logarithms, E is the activation energy, R is the gas constant (8.314 J/mole K), and T is the absolute temperature. Taking the natural log of each side:

$$\ln k = \ln A - Ea/RT.$$

For a reaction that is three times as fast, the Arrhenius equation becomes

$$3k = Ae^{-E_a/R(T+10°)}.$$

Taking the natural log:

$$\ln 3 + \ln k = \ln A - E_a / R(T + 10°).$$

Subtracting the equation for the final state from the equation for the initial state:

$$\ln k = \ln A - E_a / RT$$
$$-\left(\ln 3 + \ln k = \ln A - E_a / R(T + 10)\right)$$
$$\overline{}$$
$$-\ln 3 = -E_a / RT + E_a / R(T + 10)$$

Solving for E_a:

$$-\ln 3 = -E_a/RT + E_a/R(T + 10) \qquad R = 8.314 \text{ J/mole K}$$
$$T = 0 + 273 = 273 \text{ K}$$

$$-\ln 3 = -E_a/(8.314 \text{ J/mole K})(273 \text{ K}) + E_a/(8.314 \text{ J/mole K})$$
$$(283 \text{ K})$$

$$- 1.10 = -E_a/2{,}269.72 \text{ J/mole} + E_a/2{,}352.86 \text{ J/mole}$$

$$(2{,}269.72 \text{ J/mole})(2{,}352.86 \text{ J/mole}) \times -1.10 =$$
$$(-E_a)(2{,}352.86 \text{ J/mole})$$
$$+ E_a (2{,}269.72 \text{ J/mole})$$

$$- 5.874 \times 10^6 \text{ J}^2/\text{mole}^2 = -8.314 \times 10^1 \text{ J/mole} \times E_a$$
$$7.07 \times 10^4 \text{ J/mole} = E_a$$

6.5   Activation Energy

The activation energy is the energy necessary to cause a reaction to occur. It is equal to the difference in energy between the transition state (or "activated complex") and the reactants.

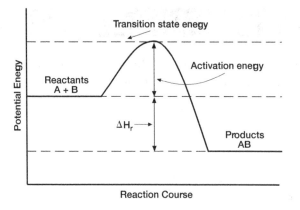

Reaction Course

$$\Delta E = \Sigma E \text{ products} - \Sigma E \text{ reactants}$$

In an exothermic process, energy is released and ΔE of reaction is negative; in an endothermic process, energy is absorbed and ΔE is positive.

For a reversible reaction, the energy liberated in the exothermic reaction equals the energy absorbed in the endothermic reaction. (The energy of the reaction, ΔE, is equal also to the difference between the activation energies of the opposing reactions, $\Delta E = E_a - E_{a'}$.)

A catalyst affects a chemical reaction by lowering the activation energy equally for both the forward and the reverse reactions.

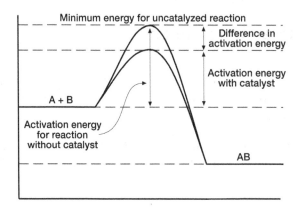

Problem Solving Example:

Q For the system described by the following diagram, is the forward reaction exothermic or endothermic?

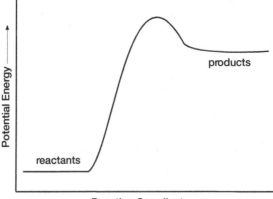

A By observing the diagram, we see that the products are at a higher potential energy than the reactants, which means that energy must have been absorbed during the course of the reaction. The only mechanism by which this can occur is absorption of heat and subsequent conversion of heat into potential energy. Hence, the reaction absorbs heat and is therefore endothermic.

6.6 Reaction Rate Law

The rate of an irreversible reaction is directly proportional to the concentration of the reactants raised to some power. For the reaction $A + B \rightarrow$ products, the rate $\propto [A]^x[B]^y$.

The order of the reaction with respect to A is x, the order with respect to B is y, and the overall order (sum of the individual orders) is $x + y$.

The following equation is termed the rate law for the reaction:

$$\text{rate} = k[A]^x[B]^y$$

where k is the rate constant.

First order reactions:

$$\text{rate} = k[A]$$

If the reaction rate is doubled by doubling the concentration of the reactant, the order with respect to the reactant is 1.

Second order reactions:

$$\text{rate} = k[A]^2$$

$$\text{rate} = k[2a]^2 = 4\,ka^2 \text{ (Effect of increasing } [A] \text{ from } a \text{ to } 2a)$$

If the rate is increased by a factor of four when the concentration of a reactant is doubled, the reaction is second order with respect to that component.

Third order reactions:

The rate of a third order reaction would undergo an eight-fold increase when the concentration is doubled ($2^3 = 8$).

Problem Solving Example:

 Assume that an A molecule reacts with two B molecules in a one-step process to give AB_2. (a) Write a rate law for this reaction. (b) If the initial rate of formation of AB_2 is 2.0×10^{-5} M/s and the initial concentrations of A and B are 0.30 M each, what is the value of the specific rate constant? (c) What is the overall order for this rate equation?

(a) The overall equation for this reaction is

$$A + 2B \rightarrow AB_2 .$$

Since no other information is provided about the reaction, the rate law for the reaction is assumed to be written

$$\text{rate} = k\,[A][B]^2,$$

where k is the rate constant and [ ] indicates concentration.

(b) One can solve for k using the rate law when the rates, [A] and [B], are given as they are in this problem.

$$\text{rate} = k\,[A][B]^2$$

$$2.0 \times 10^{-5}\,M/s = k(0.30\,M)(0.30\,M)^2$$

$$\frac{2.0 \times 10^{-5}\,M/s}{(0.30M)(0.30M)^2} = k$$

$$k = 7.4 \times 10^{-4}\,M^{-2}\,s^{-1}$$

(c) The overall order of a rate equation is equal to the sum of the exponents to which the concentrations are raised. In the equation

$$\text{rate} = k\,[A][B]^2,$$

[B] is raised to the second power and [A] is raised to the first power. Hence, the rate is second order in [B], first order in [A], and $2 + 1 = 3$, or third order overall.

6.7   Collision Theory of Reaction Rates

The rate of reaction depends on two factors: the number of collisions per unit time, and the fraction of these collisions that result in a reaction.

The following diagrams show that the number of collisions, and consequently, the rate of reaction, is proportional to the product of the concentrations. The rate of the reaction is directly proportional to the concentration.

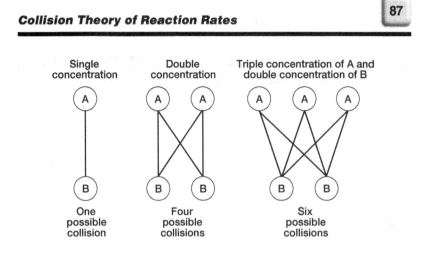

Single concentration	Double concentration	Triple concentration of A and double concentration of B
One possible collision	Four possible collisions	Six possible collisions

The rate of a reaction according to collision theory is

$$\text{rate} = f \times Z,$$

where Z is the total number of collisions and f is the fraction of the total number of collisions occurring at a sufficiently high energy for reaction. The rate of the reaction is further decreased by a factor, p, which is a measure of the importance of the molecular orientations during collision:

$$\text{rate} = pfZ.$$

Z, the collision frequency, is proportional to the concentrations of the reacting molecules:

$$Z = Z_0[A]^n[B]^m$$

where Z_0 is the collision frequency when all of the reactants are at unit concentration.

Therefore, $\text{rate} = pfZ_0[A]^n[B]^m$, or

$\text{rate} = k[A]^n[B]^m$, where $k = pfZ_0$.

Problem Solving Example:

 For the reaction $A + 2B \rightarrow C$, when all the reactants are at unit concentration, A and B collide with a frequency, Z_0, of 0.6 M/s with 1/10 collisions resulting in the formation of C. If molecular orientation, p, is assumed to be unimportant, what is the rate of the reaction when $[A] = 0.1\ M$ and $[B] = 0.2\ M$?

 $$\text{rate} = pfZ_0[A]^n[B]^m$$

Since molecular orientation is unimportant, the equation becomes rate $= fZ_0[A]^n[B]^m$, where f is the fraction of the total number of collisions occurring at a sufficiently high energy for reaction to occur.

Therefore, rate $= (0.1)(0.6\ M/\text{s})(0.1\ M)(0.2\ M)^2$

$$= 2 \times 10^{-4}\ M^4/\text{s}.$$

Quiz: Chemical Formulas—Types and Rates of Chemical Reactions

1. $HCl + NaOH \rightarrow NaCl + H_2O$ is an example of a reaction classified as

    (A) decomposition.

    (B) double replacement.

    (C) reversible.

    (D) single replacement.

    (E) synthesis.

2. The name of the compound $HClO_2$ is

    (A) hydrochloric acid.

    (B) hypochlorous acid.

    (C) chlorous acid.

    (D) chloric acid.

    (E) perchloric acid.

3. Which is the empirical formula of a compound consisting of 70% iron and 30% oxygen?

    (A) FeO   (D) Fe_2O_3

    (B) FeO_2 (E) Fe_3O_5

    (C) Fe_2O

4. For the reaction $2A + B \rightarrow 2C$, the following rate data was obtained:

Experiment	initial conc. of A mol/L	initial conc. of B mol/L	initial rate mol/L s
1	0.5	0.5	10
2	0.5	1.0	20
3	0.5	1.5	30
4	1.0	0.5	40

Which expression below states the rate law for the reaction above?

(A) Rate = $k[A]^2[B]$ (D) Rate = $k[A]^3$

(B) Rate = $k[A][B]$ (E) Rate = $k[A]^2[B]^3$

(C) Rate = $k[A]^2[B]^2$

5. In a lab, each of the following factors will vary to affect reaction rate EXCEPT

(A) catalyst used.

(B) concentration of reactants.

(C) identity of reactants.

(D) quantity of reactants.

(E) temperature.

6. The activation energy of a reaction can be determined from the slope of which of the following graphs?

(A) $\ln k$ vs T (D) $\dfrac{T}{\ln k}$ vs $\dfrac{1}{T}$

(B) $\dfrac{\ln k}{T}$ vs $\dfrac{1}{T}$ (E) $\dfrac{\ln k}{1}$ vs $\dfrac{1}{T}$

(C) $\ln k$ vs $\dfrac{1}{T}$

7. Which of the following depicts an endothermic reaction (abscissa = reaction time; ordinate = energy)?

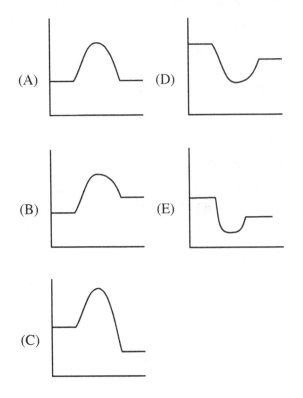

(A)   (D)

(B)   (E)

(C)

8. For a certain second order reaction $A + B \rightarrow C$, it was noted that when the initial concentration of A is doubled while B is held constant, the initial reaction rate doubles, and when the initial concentration of B is doubled while A is held constant, the initial reaction rate increases four-fold. What is the rate expression for this reaction?

(A) $r = k[A][B]$ (D) $r = k[A]^3[B]$

(B) $r = k[A]^2[B]^3$ (E) $r = k[A]^2[B]$

(C) $r = k[A][B]^2$

9. The compound $CuSO_4$ is named

    (A) cuprous sulfide.   (D) copper (II) sulfate.

    (B) cupric sulfide.   (E) sulfurous copper.

    (C) copper tetrasulfate.

10. The correct formula for the compound zinc chloride is

    (A) $ZnCl$.   (D) Zn_2Cl.

    (B) $ZnCl_2$.   (E) Zn_2Cl_3.

    (C) $ZnCl_3$.

ANSWER KEY

1.	(B)		6.	(C)
2.	(C)		7.	(B)
3.	(D)		8.	(C)
4.	(A)		9.	(D)
5.	(D)		10.	(B)

CHAPTER 7

Gases

7.1 Volume and Pressure

A gas has no shape of its own; rather, it takes the shape of its container. It has no fixed volume, but is compressed or expanded as its container changes in size. The volume of a gas is the volume of the container in which it is held.

Pressure is defined as force per unit area. Atmospheric pressure is measured using a barometer.

Atmospheric pressure is directly related to the length (h) of the column of mercury in a barometer and is expressed in mm or cm of mercury (Hg).

Standard atmospheric pressure is expressed in several ways: 14.7 pounds per square inch (psi), 760 mm of Hg, 760 torr, or simply 1 "atmosphere" (1 atm).

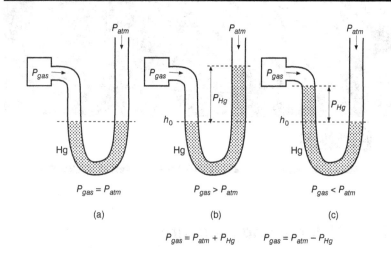

$P_{gas} = P_{atm}$

(a)

$P_{gas} > P_{atm}$

(b)

$P_{gas} = P_{atm} + P_{Hg}$

$P_{gas} < P_{atm}$

(c)

$P_{gas} = P_{atm} - P_{Hg}$

Open-end manometer

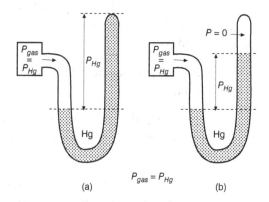

(a)

$P_{gas} = P_{Hg}$

(b)

Closed-end manometer

Problem Solving Example:

Q Consider the manometer, illustrated on the following page, first constructed by Robert Boyle. When $h = 40$ mm, what is the pressure of the gas trapped in the volume, V_{gas}? The temperature is constant, and atmospheric pressure is $P_{atm} = 1$ atm.

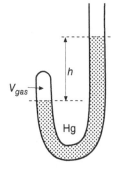

 We do not need to know any gas law to solve this problem. All we must realize is that the pressure exerted on the gas, P_{total}, is equal to the sum of the pressure exerted by the mercury, P_{Hg}, and the pressure exerted by the air, P_{atm}. Since 1 mm Hg = 1 torr and 1 atm = 760 torr,

P_{Hg} = 40 mm Hg = 40 torr and P_{atm} = 1 atm = 760 torr.

Then $P_{total} = P_{Hg} + P_{atm}$ = 40 torr + 760 torr = 800 torr.

7.2 Boyle's Law

Boyle's law states that, at a constant temperature, the volume of a gas is inversely proportional to the pressure:

$$V \propto \frac{1}{P} \quad \text{or } V = \text{constant} \times \frac{1}{P} \text{ or } PV = \text{constant}.$$

$$P_1 V_1 = P_2 V_2$$

$$V_2 = V_1 \left(\frac{P_1}{P_2} \right)$$

A hypothetical gas that would follow Boyle's law under all conditions is called an ideal gas. Deviations from Boyle's law that occur with real gases represent nonideal behavior.

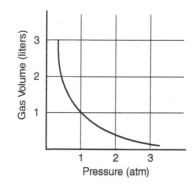

Problem Solving Example:

 What pressure is required to compress 5 L of gas at 1 atm pressure to 1 L at a constant temperature?

In solving this problem, one uses Boyle's law: the volume of a given mass of gas at constant temperature varies inversely with the pressure. This means that, for a given gas, the pressure and the volume are proportional at a constant temperature and their product equals a constant.

$$P \times V = K$$

where P is the pressure, V is the volume, and K is a constant. From this one can propose the following equation:

$$P_1 V_1 = P_2 V_2$$

where P_1 is the original pressure, V_1 is the original volume, P_2 is the new pressure, and V_2 is the new volume.

In this problem, one is asked to find the new pressure and is given the original pressure and volume and the new volume.

$$P_1V_1 = P_2V_2$$

$$1 \text{ atm} \times 5 \text{ L} = P_2 \times 1 \text{ L}$$

$$\frac{1 \text{ atm} \times 5 \text{ L}}{1 \text{ L}} = P_2$$

$$5 \text{ atm} = P_2$$

$$P_1 = 1 \text{ atm}$$
$$V_1 = 5 \text{ L}$$
$$P_2 = ?$$
$$V_2 = 1 \text{ L}$$

7.3   Charles' Law

Charles' law states that at constant pressure, the volume of a given quantity of a gas varies directly with the temperature:

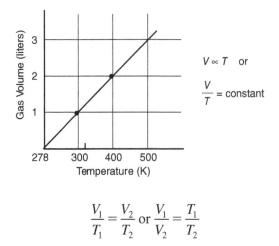

$V \propto T$ or

$\dfrac{V}{T}$ = constant

$$\frac{V_1}{T_1} = \frac{V_2}{T_2} \text{ or } \frac{V_1}{V_2} = \frac{T_1}{T_2}$$

If Charles' law were strictly obeyed, gases would not condense when they are cooled. This means that gases behave in an ideal fashion only at relatively high temperatures and low pressures.

Problem Solving Example:

Q A certain gas occupies a volume of 100 mL at a temperature of 20 °C. What will its volume be at 10 °C, if the pressure remains constant?

A In a gaseous system, when the volume is changed by increasing the temperature and the pressure is kept constant, Charles' law can be used to determine the new volume. Charles' law states that, at a constant pressure, the volume of a given mass of gas is directly proportional to the absolute temperature. Charles' law may also be written

$$\frac{V_1}{T_1} = \frac{V_2}{T_2}$$

where V_1 is the volume at the original temperature, T_1, and V_2 is the volume at the new temperature, T_2.

To use Charles' law, the temperature must be expressed on the absolute scale. The absolute temperature is calculated by adding 273 to the temperature in degrees celsius. In this problem, the celsius temperatures are given and one must convert them to the absolute scale.

$$T_1 = 20 \text{ °C} + 273 = 293 \text{ K}$$
$$T_2 = 10 \text{ °C} + 273 = 283 \text{ K}$$

Using Charles' law,

$$V_1 = 100 \text{ mL}$$
$$T_1 = 293 \text{ K}$$
$$T_2 = 283 \text{ K}$$
$$V_2 = ?$$

$$\frac{V_1}{T_1} = \frac{V_2}{T_2} \qquad V_2 = \frac{V_1 T_2}{T_1}$$

$$V_2 = \frac{(100 \text{ mL})(283 \text{ K})}{293 \text{ K}}$$

$$V_2 = 96.6 \text{ mL}$$

7.4   Dalton's Law of Partial Pressures

The pressure exerted by each gas in a mixture is called its partial pressure. The total pressure exerted by a mixture of gases is equal to the sum of the partial pressures of the gases in the mixture. This statement, known as Dalton's law of partial pressures, can be expressed

$$P_T = P_a + P_b + P_c + \ldots.$$

When a gas is collected over water (a typical laboratory method), some water vapor mixes with the gas. The total gas pressure then is given by

$$P_T = P_{gas} + P_{H_2O},$$

where P_{gas} = pressure of dry gas and P_{H_2O} = vapor pressure of water at the temperature of the system.

Problem Solving Example:

 A mixture of nitrogen and oxygen gas is collected by displacement of water at 30 °C and 700 torr pressure. If the partial pressure of nitrogen is 550 torr, what is the partial pressure of oxygen? (Vapor pressure of H_2O at 30 °C = 32 torr.)

 Here, one uses Dalton's law of partial pressures. This law can be stated: each of the gases in a gaseous mixture behaves independently of the other gases and exerts its own pressure. The total pressure of the mixture is the sum of the partial pressures exerted by each gas present. Stated algebraically,

$$P_T = P_a + P_b + P_c + \ldots,$$

where P_T is the total pressure and P_a, P_b, P_c ... are the partial pressures of the gases present. In this problem, one is told that oxygen and nitrogen are collected over H_2O, which means that there will also be water vapor present in the gaseous mixture. In this case, the equation for the total pressure can be written

$$P_T = P_{O_2} + P_{N_2} + P_{H_2O}.$$

One is given P_T, P_{N_2}, and P_{H_2O} and asked to find P_{O_2}. This can be done by using the law of partial pressures.

$P_T = P_{O_2} + P_{N_2} + P_{H_2O}$ $P_T = 700$ torr

          $P_{O_2} = ?$

700 torr $= P_{O_2} + 550$ torr $+ 32$ torr $P_{N_2} = 550$ torr

$P_{O_2} = (700 - 550 - 32)$ torr $P_{H_2O} = 32$ torr

$P_{O_2} = 118$ torr

7.5 Law of Gay-Lussac

The law of Gay-Lussac states that at constant volume, the pressure exerted by a given mass of gas varies directly with the absolute temperature:

$P \propto T$ (where volume and mass of gas are constant).

$$\frac{P_1}{T_1} = \frac{P_2}{T_2}$$

Gay-Lussac's law of combining volumes states that when reactions take place in the gaseous state, under conditions of constant temperature and pressure, the volumes of reactants and products can be expressed as ratios of small whole numbers.

Problem Solving Example:

 The air in a tank has a pressure 640 mm Hg at 23 °C. When placed in sunlight the temperature rose to 48 °C. What was the pressure in the tank?

 The law of Gay-Lussac deals with the relationship existing between pressure and the absolute temperature (°C + 273) for a given mass of gas at constant volume. The relationship is expressed in the law of Gay-Lussac: volume constant, the pressure exerted by a given mass of gas varies directly with the absolute temperature. That is:

$$P \propto T \text{ (volume and mass of gas constant)}.$$

The variation that exists between pressure and temperature at different states can be expressed as

$$\frac{P_1}{T_1} = \frac{P_2}{T_2}$$

where P_1 = pressure of original state, T_1 = absolute temperature of original state, P_2 = pressure of final state, and T_2 = absolute temperature of final state.

Thus, this problem is solved by substituting the given values into Gay-Lussac's law.

$P_1 = 640 \text{ mm Hg}$ $T_1 = 23 \text{ °C} + 273 = 296 \text{ K}$

$P_2 = ?$ $T_2 = 48 \text{ °C} + 273 = 321 \text{ K}$

Substituting and solving,

$$\frac{640 \text{ mm Hg}}{296 \text{ K}} = \frac{P_2}{321 \text{ K}}$$

$$P_2 = 640 \text{ mm Hg} \times \frac{321 \text{ K}}{296 \text{ K}}$$

$$= 694 \text{ mm Hg}.$$

7.6   Ideal Gas Law

$$V \propto \frac{1}{P}, \quad V \propto T, \quad V \propto n,$$

where n is the number of moles of a gas, then $V \propto = \frac{nT}{P}$

$$PV = nRT \text{ where } R \text{ is a constant.}$$

The hypothetical ideal gas obeys exactly the mathematical statement of the ideal gas law. This statement is also called the equation of state of an ideal gas because it relates the variables (P, V, n, T) that specify properties of the gas. Molecules of ideal gases have no attraction for one another and have no intrinsic volume; they are "point particles." Real gases act in a less than ideal way, especially under conditions of increased pressure and/or decreased temperature. Real gas behavior approaches that of ideal gases as the gas pressure becomes very low. The ideal gas law is thus considered a "limiting law."

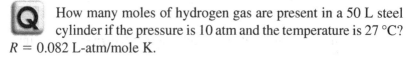

Problem Solving Example:

Q How many moles of hydrogen gas are present in a 50 L steel cylinder if the pressure is 10 atm and the temperature is 27 °C? $R = 0.082$ L-atm/mole K.

A In this problem, one is asked to find the number of moles of hydrogen gas present where the volume, pressure, and temperature are given. This would indicate that the ideal gas law should be used because this law relates these quantities to each other. The ideal gas law can be stated

$$PV = nRT,$$

where P is the pressure, V is the volume, n is the number of moles, R is the gas constant (0.082 L-atm/mole K), and T is the absolute temperature. Here, one is given the temperature in °C, which means it must be converted to the absolute scale. P, V, and R are also known. To convert a temperature in °C to the absolute scale, add 273 to the temperature in °C.

$$T = 27 \,°C + 273 = 300 \text{ K}$$

Using the ideal gas law:

$PV = nRT$ or
$P \quad = 10 \text{ atm}$
$V \quad = 50 \text{ L}$
$R \quad = 0.082 \text{ L-atm/mole K}$
$T \quad = 300 \text{ K}$
$n \quad = $ number of moles of H_2 present

$$n = \frac{(10 \text{ atm})(50 \text{ L})}{(0.082 \text{ L - atm/mole K})(300 \text{ K})}$$

$$= 20 \text{ moles.}$$

7.7   Combined Gas Law

The combined gas law states that for a given mass of gas, the volume is inversely proportional to the pressure and directly proportional to the absolute temperature. This law can be written

$$\frac{P_1 V_1}{T_1} = \frac{P_2 V_2}{T_2}$$

where P_1 is the original pressure, V_1 is the original volume, T_1 is the original absolute temperature, P_2 is the new pressure, V_2 is the new volume, and T_2 is the new absolute temperature.

Problem Solving Example:

 Calculate the pressure required to compress 2 L of a gas at 700 mm Hg pressure and 20 °C into a container of 0.1 L capacity at a temperature of –150 °C.

 One is dealing with changing volumes, pressures, and temperatures of a gas. Therefore, this problem can be solved using the combined gas law. It states that as the pressure increases, the volume decreases and that as the temperature increases, the volume increases. These factors are related by the equation

$$\frac{P_1 V_1}{T_1} = \frac{P_2 V_2}{T_2}$$

where P_1, V_1, and T_1 are the initial pressure, volume, and temperature and P_2, V_2, and T_2 are the final values.

For any problem dealing with gases, the first step always involves converting all of the temperatures to the Kelvin scale by the equation

$$K = °C + 273 .$$

For this question

$$T_1 = 20 °C + 273 = 293 \text{ K}$$
$$T_2 = -150 °C + 273 = 123 \text{ K}.$$

This seems to indicate that the pressure would decrease. But one is also told that the volume decreases, which would have the effect of increasing the pressure. Therefore, one cannot predict the final change in volume.

For the sake of clarity, set up a table as given below.

$$P_1 = 700 \text{ mm Hg} \qquad P_2 = ?$$
$$V_1 = 2 \text{ L} \qquad V_2 = 0.1 \text{ L}$$
$$T_1 = 293 \text{ K} \qquad T_2 = 123 \text{ K}$$

Since one is given five of the six values, it is possible to use the combined gas law equation to determine P_2.

$$\frac{P_1 V_1}{T_1} = \frac{P_2 V_2}{T_2}$$

$$P_2 = \frac{T_2 V_1 P_1}{T_1 V_2}$$

$$= \frac{123 \text{ K}(2 \text{ L})(700 \text{ mm Hg})}{293 \text{ K}(0.1 \text{ L})}$$

$$= 6,000 \text{ mm Hg}$$

7.8   Avogadro's Law (The Mole Concept)

Avogadro's law states that under conditions of constant temperature and pressure, equal volumes of different gases contain equal numbers of molecules.

If the initial and final pressure and temperature are the same, then the relationship between the number of molecules, N, and the volume, V, is

$$\frac{V_2}{V_1} = \frac{N_2}{V_1}.$$

The laws of Boyle, Charles, Gay-Lussac, and Avogadro are all simple corollaries of the general equation of state for an ideal gas — $PV = nRT$ — under various restraining conditions (constant T, constant P, constant V, and constants T and P, respectively; n is assumed invariant for all).

Problem Solving Example:

Q Compare the number of H_2 and N_2 molecules in two containers described as follows: (1) A 2 L container of hydrogen filled at 127 °C and 5 atm. (2) A 5 L container of nitrogen filled at 27 °C and 3 atm.

A Avogadro's law states that equal numbers of molecules are contained in equal volumes of different gases if the pressure and temperature are the same. Therefore, if the conditions were the same, then there would be equal numbers of molecules of H_2 and N_2.

For reasons of clarity, it is useful to set up a table like that shown below.

gas	V	T	P
H_2	2 L	400 K	5 atm
N_2	5 L	300 K	3 atm

By using the general gas equation

$$PV = nRT$$

where n = number of moles and R = 0.082 L-atm/mole K, one can find the number of moles of H_2 and N_2.

Thus, for H_2

$$n = \frac{PV}{RT} = \frac{(5 \text{ atm})(2 \text{ L})}{(0.082 \text{ L-atm / mole K})(400 \text{ K})} = 0.3 \text{ moles}$$

and for N_2

$$n = \frac{PV}{RT} = \frac{(3 \text{ atm})(5 \text{ L})}{(0.082 \text{ L-atm / mole K})(300 \text{ K})} = 0.6 \text{ moles.}$$

There are twice as many moles of N_2 as there are of H_2. Since the number of molecules is directly proportional to the number of moles, then there are twice as many molecules of N_2 as molecules of H_2.

7.9 Real Gases

Real gases fail to obey the ideal gas law under most conditions of temperature and pressure.

Real gases have a finite (non zero) molecular volume; i.e., they are not true "point particles." The volume within which the molecules may not move is called the excluded volume. The real volume (volume of the container) is therefore slightly larger than the ideal volume (the volume the gas would occupy if the molecules themselves occupied no space):

$$V_{real} = V_{ideal} + nb$$

where b is the excluded volume per mole and n is the number of moles of gas. The ideal pressure, that is, the pressure the gas could exert in the absence of intermolecular attractive forces, is higher than the actual pressure by an amount that is directly proportional to n^2/V^2:

$$P_{ideal} = P_{real} + \frac{n^2 a}{V^2},$$

where a is a proportionality constant that depends on the strength of the intermolecular attractions. Therefore,

$$\left(P + \frac{n^2 a}{V^2} \right)(V - nb) = nRT$$

is the van der Waal's equation of state for a real gas.

The values of the constants a and b depend on the particular gas and are tabulated for many real gases.

Problem Solving Example:

Q Using van der Waal's equation, calculate the pressure exerted by 1 mole of carbon dioxide at 0 °C in a volume of (a) 1.00 L and (b) 0.05 L.

A Real gases are more compressible than ideal gases because their molecules attract each other. The intermolecular attraction is provided for by adding to the observed pressure a term of n^2a/V^2 in the ideal gas law ($PV = nRT$), where n is the number of moles, V is the volume, and a is a van der Waal's constant. At very high pressures, real gases occupy larger volumes than ideal gases; this effect is provided for by subtracting an excluded volume nb from the observed volume V to give the actual volume in which the molecules move (b is a constant). Thus, van der Waal's full equation is:

$$\left(P + \frac{n^2a}{V^2}\right)(V - nb) = nRT.$$

To solve this problem one must: (1) convert the temperature from units of °C to K by adding 273 to the °C temperature. This is done because T is expressed in the absolute. (P = pressure and R = gas constant.) (2) Find the values for the constants (a) and (b) (these can be found in any reference text). For CO_2, a = 3.592 L² atm/mole²; b = 0.04267 L/mole. (3) Substitute the known values into the van der Waal's equation and solve for pressure.

Known values:

$$n = 1, R = 0.082 \frac{\text{L-atm}}{\text{mole K}}, T = 273 \text{ K}.$$

(a) V = 1.00 L. Substitute

$$\left(P + \frac{(1)^2(3.592)}{(1.00)^2}\right)(1.00 - (1)(0.04267)) = (1)(0.082)(273 \text{ K})$$

$$P = \frac{(0.082)(273\text{ K})}{(1.00 - 0.04267)} - \frac{3.592}{(1.00)^2}$$
$$= 23.38 - 3.592$$
$$= 20 \text{ atm}$$

(b) $V = 0.05$ L. Substitute

$$\left(P + \frac{(1)^2(3.592)}{(0.05)^2}\right)(0.05 - (1)(0.04267)) = (1)(0.082)(273\text{ K})$$

$$P = \frac{(0.082)(273\text{ K})}{(0.05 - 0.04267)} - \frac{3.592}{(0.05)^2}$$
$$= 3,054 - 1,437$$
$$= 2,000 \text{ atm}$$

7.10 Graham's Law of Effusion and Diffusion

Effusion is the process in which a gas escapes from one chamber of a vessel to another by passing through a very small opening or orifice.

Graham's law of effusion states that the rate of effusion is inversely proportional to the square root of the density of the gas:

$$\text{rate of effusion} \propto \sqrt{\frac{1}{d}}, \text{ and}$$

$$\frac{\text{rate of effusion}(A)}{\text{rate of effusion}(B)} = \sqrt{\frac{d_B}{d_A}} = \sqrt{\frac{M_B}{M_A}}$$

where M is the molecular weight of each gas, and where the temperature is the same for both gases.

Mixing of molecules of different gases by random motion and collision until the mixture becomes homogeneous is called diffusion.

Graham's law of diffusion states that the relative rates at which gases will diffuse will be inversely proportional to the square roots of their respective densities or molecular weights:

$$\text{rate} \propto \frac{1}{\sqrt{\text{mass}}} \text{(where, again, } T_1 = T_2 \text{) and}$$

$$\frac{\text{rate 1}}{\text{rate 2}} = \sqrt{\frac{M_2}{M_1}} \text{ or } \left(\frac{r_1}{r_2} = \frac{\sqrt{d_2}}{d_1} \right).$$

Problem Solving Example:

Q Two gases, HBr and CH_4, have molecular weights of 81 and 16, respectively. The HBr effuses through a certain small opening at the rate of 4 mL/s. At what rate will the CH_4 effuse through the same opening?

A The comparative rates or speeds of effusion of gases are inversely proportional to the square roots of their molecular weights. This is written

$$\frac{\text{rate}_A}{\text{rate}_B} = \frac{\sqrt{M_B}}{\sqrt{M_A}}.$$

For this case
$$\frac{\text{rate}_{HBr}}{\text{rate}_{CH_4}} = \frac{\sqrt{M_{CH_4}}}{\sqrt{M_{HBr}}}.$$

One is given the rate_{HBr}, M_{CH_4}, and M_{HBr} and asked to find rate_{CH_4}.

Solving for rate_{CH_4}:

$$\frac{\text{rate}_{HBr}}{\text{rate}_{CH_4}} = \frac{\sqrt{M_{CH_4}}}{\sqrt{M_{HBr}}}.$$

$$\frac{4 \text{ mL/s}}{\text{rate}_{CH_4}} = \frac{\sqrt{16}}{\sqrt{81}}$$

$$\text{rate}_{HBr} = 4 \text{ mL/s}.$$

$$\text{rate}_{CH_4} = ?$$

$$M_{CH_4} = 16 \text{ g/mole}$$

$$\text{rate}_{HBr} = 81 \text{ g/mole}$$

$$\text{rate}_{CH_4} = \frac{4 \text{ mL/s} \times \sqrt{81}}{\sqrt{16}}$$

$$= \frac{4 \text{ mL/s} \times 9}{4} = 9 \text{ mL/s}.$$

7.11   The Kinetic Molecular Theory

The kinetic molecular theory is summarized as follows:

1. Gases are composed of tiny, invisible molecules that are widely separated from one another in otherwise empty space.

2. The molecules are in constant, continuous, random, and straight-line motion.

3. The molecules collide with one another, but the collisions are perfectly elastic (that is, they result in no net loss of energy).

4. The pressure of a gas is the result of collisions between the gas molecules and the walls of the container.

5. The average kinetic energy of all the molecules collectively is directly proportional to the absolute temperature of the gas. The average kinetic energy of equal numbers of molecules of any gas is the same at the same temperature.

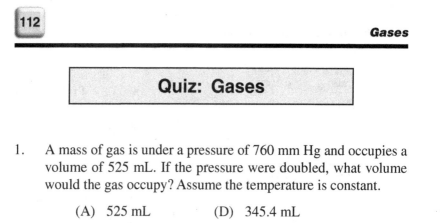

Quiz: Gases

1. A mass of gas is under a pressure of 760 mm Hg and occupies a volume of 525 mL. If the pressure were doubled, what volume would the gas occupy? Assume the temperature is constant.

   (A) 525 mL   (D) 345.4 mL

   (B) 760.1 mL   (E) 104.3 mL

   (C) 262.5 mL

2. The relationship between the absolute temperature and volume of a gas at constant pressure is given by

   (A) Boyle's law.

   (B) Charles' law.

   (C) the combined gas law.

   (D) Avogadro's law.

   (E) None of the above.

3. A sample of a pure gas occupied a volume of 500 mL at a temperature of 27 °C and a pressure of 0.4 atm. The number of moles present in this sample is

   (A) 0.045.   (D) 8.13.

   (B) 0.008.   (E) 0.182.

   (C) 0.091.

4. A gas at 43 °C has a pressure of 680 torr. If the temperature is raised to 55 °C, what will the new pressure of the gas be? (Assume all other variables remain constant.)

    (A) 735 torr   (D) 870 torr

    (B) 706 torr   (E) 374 torr

    (C) 158 torr

5. An ideal gas is most likely to be found under conditions of

    (A) low temperatures. (D) Both (B) and (C).

    (B) high temperatures. (E) high pressure.

    (C) low pressure.

6. The relationship between the pressure, volume, and absolute temperature is given by

    (A) Boyle's law.

    (B) Charles' law.

    (C) the combined gas law.

    (D) the ideal gas law.

    (E) None of the above.

7. A flask containing H_2 at 0 °C was sealed off at a pressure of 1 atm and the gas was found to weigh 0.4512 g (AW H = 1.008 g/mole). How many moles of H_2 were present?

    (A) 0.2238   (D) 4.151

    (B) 0.4413   (E) 0.1692

    (C) 2.016

8. Which of the following represents the real gas law?

   (A) $nRT = (P + a/V^2)(V - b)$

   (B) $nRT = (P - a/V^2)(V - b)$

   (C) $nRT = (P + a/V^2)(V - nb)$

   (D) $nRT = (P - a/V^2)(V + nb)$

   (E) $nRT = (P + n^2a/V^2)(V - nb)$

9. The rate of diffusion of hydrogen gas as compared to that of oxygen gas is

   (A) half as fast.

   (B) identical.

   (C) twice as fast.

   (D) four times as fast.

   (E) eight times as fast.

10. Select the characteristic that is NOT a standard condition for comparing gas volumes.

    (A) Pressure: 31 inches

    (B) Pressure: 760 torr

    (C) Temperature: 0° Centigrade

    (D) Temperature: 32° Fahrenheit

    (E) Temperature: 273 Kelvin

ANSWER KEY

1.	(C)	6.	(C)
2.	(B)	7.	(A)
3.	(B)	8.	(E)
4.	(B)	9.	(D)
5.	(D)	10.	(A)

CHAPTER 8

Liquids, Solids, and Phase Changes

8.1 Liquids

A liquid is composed of molecules that are constantly and randomly moving.

8.1.1 Volume and Shape

Liquids maintain a definite volume, but because of their ability to flow, their shape depends on the contour of the container holding them.

8.1.2 Compression and Expansion

In a liquid the attractive forces hold the molecules close together, so that increasing the pressure has little effect on the volume. Therefore, liquids are incompressible. Changes in temperature cause only small volume changes.

8.1.3   Diffusion

Liquids diffuse much more slowly than gases because of the constant interruptions in the short mean free paths between molecules.

The rates of diffusion in liquids are more rapid at higher temperatures.

8.1.4   Surface Tension

The strength of the inward forces of a liquid is called the liquid's surface tension. Surface tension decreases as the temperature increases.

8.1.5   Kinetics of Liquids

Increases in temperature increase the average kinetic energy of molecules and the rapidity of their movement. If a particular molecule gains enough kinetic energy when it is near the surface of a liquid, it can overcome the attractive forces of the liquid phase and escape into the gaseous phase. This is called a change of phase (specifically, evaporation).

8.2   Heat of Vaporization and Heat of Fusion

The heat of vaporization of a substance is the number of calories required to convert 1 g of liquid to 1 g of vapor without a change in temperature.

The reverse process, changing 1 g of gas into a liquid without change in temperature, requires the removal of the same amount of heat energy (the heat of condensation).

The heat needed to vaporize 1 mole of a substance is called the molar heat of vaporization or the molar enthalpy of vaporization, $\Delta H v_{ap}$, which is also represented as

$$\Delta H_{\text{vaporization}} = H_{\text{vapor}} - H_{\text{liquid}}.$$

The magnitude of ΔH_{vap} provides a good measure of the strengths of the attractive forces operative in a liquid.

The number of calories needed to change 1 g of a solid substance (at the melting point) to 1 g of liquid (at the melting point) is called the heat of fusion.

The total amount of heat that must be removed in order to freeze 1 mole of a liquid is called its molar heat of crystallization. The molar heat of fusion, ΔH_{fus}, is equal in magnitude but opposite in sign to the molar heat of crystallization and is defined as the amount of heat that must be supplied to melt 1 mole of a solid:

$$\Delta H_{\text{fus}} = H_{\text{liquid}} - H_{\text{solid}}.$$

Problem Solving Example:

Q What weight of ice could be melted at 0 °C by the heat liberated by condensing 100 g of steam at 100 °C to liquid? Heat of vaporization = 540 cal/g and heat of fusion = 80 cal/g.

A The quantity of heat necessary to convert 1 g of a liquid into a vapor is termed the heat of vaporization. For water, 540 calories are necessary to change liquid water at 100 °C into vapor at 100 °C. In this problem vapor is condensed, thus 540 cal of heat are evolved for each gram of liquid condensed.

number of cal evolved in condensation =
    540 cal/g × weight of vapor

Here 100 g of vapor are condensed. Thus,

number of cal evolved = 540 cal/g × 100 g = 54,000 cal.

When ice melts, heat is absorbed. About 80 calories of heat are required to melt 1 g of ice, the heat of fusion. Here, 54,000 cal are evolved in the condensation. Therefore, this is the amount of heat available to melt the ice. Because 80 cal are needed to melt 1 g of ice, one can find the number of grams of ice that can be melted by 54,000 cal, by dividing 54,000 cal by 80 cal/g.

$$\text{number of grams of ice melted} = \frac{54,000 \text{ cal}}{80 \text{ cal/g}} = 700 \text{ g}$$

8.3   Raoult's Law and Vapor Pressure

When the rate of evaporation equals the rate of condensation, the system is in equilibrium.

The vapor pressure is the pressure exerted by the gas molecules when they are in equilibrium with the liquid.

The vapor pressure increases with increasing temperature.

Raoult's law states that the vapor pressure of a solution at a particular temperature is equal to the mole fraction of the solvent in the liquid phase multiplied by the vapor pressure of the pure solvent at the same temperature:

$$P_{\text{solution}} = X_{\text{solvent}} P_{\text{solvent}}^{\circ}$$

and $P_A = X_A P_A^{\circ}$, where PA is the vapor pressure of A with solute added, is the vapor pressure of pure A, and XA is the mole fraction of A in the solution. The solute is assumed here to be nonvolatile (i.e., NaCl or sucrose in water).

Problem Solving Example:

Q The vapor pressures of pure benzene and toluene at 60 °C are 385 and 139 torr, respectively. Calculate (a) the partial pressures of benzene and toluene, (b) the total vapor pressure of the solution, and (c) the mole fraction of toluene in the vapor above a solution with 0.60 mole fraction toluene.

A The vapor pressure of benzene over solutions of benzene and toluene is directly proportional to the mole fraction of benzene in the solution. The vapor pressure of pure benzene is the proportionality constant. This is analogous to the vapor pressure of toluene. This is known as Raoult's law. It may be written as

$$P_A = X_A P_A^\circ$$
$$P_B = X_B P_B^\circ$$

where A and B refer to components A and B, P_A and P_B represent the partial vapor pressure above the solution, and are the vapor pressures of pure components, and X_A and X_B are their mole fractions. Solutions are called ideal if they obey Raoult's law.

The mole fraction of a component in the vapor is equal to its pressure fraction in the vapor. The total vapor pressure is the sum of the vapor's component partial pressures.

To solve this problem one must

1. calculate the partial pressures of benzene and toluene using Raoult's law;

2. find the total vapor pressure of the solution by adding the partial pressures; and

3. find the mole fraction of toluene in the vapor.

One knows the mole fraction of toluene in the solution is 0.60 and, thus, one also knows the mole fraction of benzene is (1 − 0.60), or 0.40. Using Raoult's law:

$$P^\circ_{benzene} = 385 \text{ torr} \qquad P^\circ_{toluene} = 139 \text{ torr}$$

(a) $P_{benzene} = (0.40)(385 \text{ torr}) = 150 \text{ torr}$

$P_{toluene} = (0.60)(139 \text{ torr}) = 83 \text{ torr}$

(b) $P_{total} = 15.0 + 83 = 230 \text{ torr}$

(c) The mole fraction of toluene in the vapor =

$$X_{\text{toluene, vap}} = \frac{P_{\text{toluene}}}{P_{\text{toluene}} + P_{\text{benzene}}} = \frac{83}{230} = 0.36.$$

8.4 Boiling Point and Melting Point

The boiling point (b.p.) of a liquid is the temperature at which the pressure of vapor escaping from the liquid equals atmospheric pressure. The normal boiling point of a liquid is the temperature at which its vapor pressure is 760 mm Hg, that is, standard atmospheric pressure. Liquids with relatively strong attractive forces have high boiling points. The melting point (m.p.) of a substance is the temperature at which its solid and liquid phases are in equilibrium.

8.5 Solids

Properties of solids are as follows:

1. They retain their shape and volume when transferred from one container to another.

2. They are virtually incompressible.

3. They exhibit extremely slow rates of diffusion.

In a solid, the attractive forces between the atoms, molecules, or ions are relatively strong. The particles are held in a rigid structural array, wherein they exhibit only vibrational motion.

There are two types of solids, amorphous and crystalline. Crystalline solids are species composed of structural units bounded by specific (regular) geometric patterns. They are characterized by sharp melting points.

Amorphous substances do not display geometric regularity in the solid; glass is an example of an amorphous solid. Amorphous substances have no sharp melting point, but melt over a wide range of temperatures.

When solids are heated at certain pressures, some solids vaporize directly without passing through the liquid phase. This is called sublimation. The heat required to change 1 mole of solid A completely to vapor is called the molar heat of sublimation, ΔH_{sub}. Note that

$$\Delta H_{sub} = \Delta H_{fus} + \Delta H_{vap}.$$

Problem Solving Example:

Q When heated under pressure, 1 mole of solid A is completely vaporized. What is the heat of sublimation if the heat of fusion, ΔH_{fus}, is equal to 65 cal/g and the heat of vaporization, ΔH_{vfp}, is equal to 297 cal/g?

A When 1 mole of solid is completely vaporized under pressure, it is known as sublimation and

$$\Delta H_{sub} = \Delta H_{fus} + \Delta H_{vap}$$

$$\Delta H_{sub} = 65 \text{ cal/g} + 297 \text{ cal/g} = -362 \text{ cal/g}.$$

8.6 Phase Diagram of Water

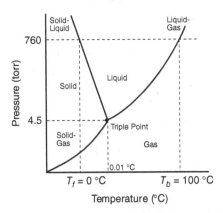

Problem Solving Example:

Q Draw a labeled phase diagram for a substance Z that has the
following properties: normal boiling point = 220 °C, normal
freezing point = 80 °C, and triple point = 60 °C and 0.20 atm. Predict
the freezing and boiling points, if the pressure was 0.80 atm.

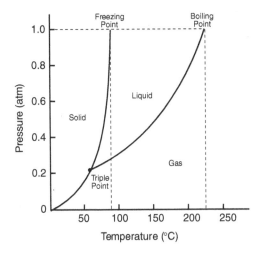

A To draw this diagram you want to understand all the terms involved. The relation between solid, liquid, and gaseous states as a function of the given temperature and pressure can be summarized on a graph known as a phase diagram. From the given experimental observations, you can draw the diagram. The lines that separate the states in a diagram represent an equilibrium between the phases. The intersection of the three lines is called the triple point, where all three phases are in equilibrium with each other. Normal boiling and melting points are those readings taken at 1 atm. Thus, the phase diagram can be written as shown in the accompanying figure.

From the diagram you see that if the pressure was 0.80 atm, the b.p. and f.p. would drop, respectively, to 205 °C and 75 °C.

8.7 Phase Equilibrium

In a closed system, when the rates of evaporation and condensation are equal, the system is in phase equilibrium.

In a closed system, when opposing changes are taking place at equal rates, the system is said to be in dynamic equilibrium. Virtually all of the equilibria considered in this review are dynamic equilibria.

Quiz: Liquids, Solids, and Phase Changes

1. A solution is made by combining 1 mole of ethanol and 2 moles of water. What is the total vapor pressure above the solution? (At the same temperature, the vapor pressure of pure ethanol is 0.53 atm and the vapor pressure of water is 0.24 atm.)

   (A) 0.34 atm   (D) 0.56 atm

   (B) 0.41 atm   (E) 0.77 atm

   (C) 0.43 atm

2. The melting point of a substance is

   (A) always 0 °C.

   (B) very low for solids.

   (C) the same as its freezing point.

   (D) very high for liquids.

   (E) −32 °C for water.

3. Which one of the following processes indicates sublimation?

   (A) gas → liquid   (D) solid → gas

   (B) gas → solid   (E) liquid → gas

   (C) solid → liquid

4. At which point can all three phases coexist at equilibrium?

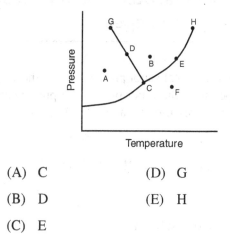

(A) C   (D) G

(B) D   (E) H

(C) E

5. Forty grams of ice at 0 °C is mixed with 100 g of water at 60 °C. What is the final temperature after equilibrium has been established? Heat of fusion of H_2O = 80 cal/g and specific heat = 1 cal/g °C.

(A) 40 °C   (D) 80 °C

(B) 20 °C   (E) 100 °C

(C) 60 °C

6. The attractive force between the protons of one molecule and the electrons of another molecule are strongest

(A) in the solid phase.

(B) in the liquid phase.

(C) in the gas phase.

(D) during sublimation.

(E) during fusion.

7. For a liquid, increasing pressure

   (A) increases volume.

   (B) greatly decreases volume.

   (C) has little effect on volume.

   (D) decreases mass.

   (E) None of the above.

8. Liquids diffuse

   (A) more slowly than gases.

   (B) more rapidly than gases.

   (C) at the same rate as gases.

   (D) more rapidly at low temperature.

   (E) None of the above.

9. Increasing the temperature of a liquid

   (A) largely increases volume.

   (B) increases surface tension.

   (C) decreases average kinetic energy.

   (D) can initiate a change in phase.

   (E) None of the above.

10. In a closed system, the system is said to be in phase equilibrium when

(A) the rate of evaporation > condensation.

(B) the rate of evaporation < condensation.

(C) the rate of evaporation = condensation.

(D) opposing changes are taking place at different rates.

(E) None of the above.

ANSWER KEY

1.	(A)	6.	(A)
2.	(C)	7.	(C)
3.	(D)	8.	(A)
4.	(A)	9.	(D)
5.	(B)	10.	(C)

CHAPTER 9

Properties of Solutions

9.1   Types of Solutions

There are three types of solutions: gaseous, liquid, and solid.

The most common type of solution consists of a solute dissolved in a liquid.

Air is an example of a gaseous solution.

Solid solutions, of which many alloys (mixtures of metals) are examples, are of two types:

1. Substitutional solid solutions in which atoms, molecules, or ions of one substance take the place of particles of another substance in its crystalline lattice.

2. Interstitial solid solutions are formed by placing atoms of one kind into voids, or interstices, that exist between atoms in the host lattice.

9.2 Concentration Units

Mole fraction is the number of moles of a particular component of a solution divided by the total number of moles of all the substances present in the solution:

$$X_A = \frac{n_A}{n_A + n_B + n_C + \ldots}$$

$$\sum_{i=1}^{N} X_i = 1.$$

Mole percent is equal to 100% × mole fraction. Weight fraction specifies the fraction of the total weight of a solution that is contributed by a particular component. Weight percent is equal to 100% × weight fraction.

Molarity (M) of a solution is the number of moles of solute per liter of solution:

$$\text{molarity } (M) = \frac{\text{moles of solute}}{\text{liters of solution}}.$$

Normality (N) of a solution is the number of equivalents of solute per liter of solution:

$$\text{normality } (N) = \frac{\text{equiv of solute}}{\text{liters of solution}}.$$

Molality of a solution is the number of moles of solute per kilogram (1,000 g) of solvent.

Problem Solving Example:

 A solution is prepared by dissolving 464 g of NaOH in water and then diluting to 1 L. The density of the resulting solution is 1.37 g/mL. Express the concentration of NaOH as (a) percentage by weight, (b) molarity, (c) normality, (d) molality, and (e) mole fraction.

A (a) The percentage by weight of NaOH in this solution is found by dividing the weight of NaOH present, 464 g, by the weight of the solution and multiplying by 100. The weight of the solution is found by using the density of the solution. The density, 1.37 g/mL, tells the weight of 1 mL of solution, namely, 1.37 g. In 1 L, there are 1,000 mL, thus the weight of the solution is 1,000 times the weight of 1 mL.

weight of 1 L = 1,000 mL $\times$ 1.37 g/mL = 1,370 g

The percentage by weight of the NaOH in the solution can now be found.

$$\text{percentage of NaOH} = \frac{\text{weight of NaOH}}{\text{weight of solution}} \times 100$$

$$= \frac{464 \text{ g}}{1,370 \text{ g}} \times 100 = 33.9\%$$

(b) The molarity is defined as the number of moles in 1 liter of solution. The molarity in this case can be found by determining the number of moles in 464 g of NaOH, which is the amount of NaOH in 1 L. The number of moles can be found by dividing 464 g by the molecular weight of NaOH. (MW of NaOH = 40.0 g/mole.)

$$\text{number of moles} = \frac{464 \text{ g}}{\text{MV}} = \frac{464 \text{ g}}{40.0 \text{ g/mole}} = 11.6 \text{ moles}$$

The molarity of this solution is thus 11.6 M.

(c) The normality of a basic solution is the number of moles of ionizable OH^- ions in 1 liter of solution. There is one ionizable OH^- ion in each NaOH, as shown by the equation:

$$NaOH \underset{\leftarrow}{\overset{\rightarrow}{\rule{0pt}{1.2em}}} Na^+ + OH^-.$$

Therefore, there are the same number of OH^- ions in the solution as NaOH molecules dissolved. Thus, the molarity equals the normality.

normality = molarity $\times$ 1 ionizable OH^-
normality = 11.6 N

(d) The molality is defined as the number of moles present in 1 kg of solvent. One has already found that the solution weighs 1,370 g; the NaOH weighs 464 g; thus the weight of the water is equal to the difference of these two figures.

weight of H_2O = weight of solution $-$ weight of NaOH
= 1,370 g $-$ 464 g = 906 g, or 0.906 kg

There are 11.6 moles of NaOH present. Therefore, the molality can now be found.

$$molality = \frac{\text{number of moles present}}{\text{number of kg present}}$$

$$= \frac{11.6 \text{ moles solute}}{0.906 \text{ kg solvent}} = 12.8 \text{ m}$$

(e) The mole fraction is equal to the number of moles of each component divided by the total number of moles in the system. The components in this system are H_2O and NaOH. One already has found that there are 11.6 moles of NaOH present, but not the number of moles of H_2O. This can be found by dividing the weight of the water by its molecular weight, 18.0 g/mole.

$$\text{number of moles} = \frac{906 \text{ g}}{18 \text{ g/moles}} = 50.3 \text{ moles}$$

The total number of moles in the system is the sum of the number of moles of H_2O and of NaOH.

$$\text{number of moles in system} = \text{moles } H_2O + \text{moles NaOH}$$
$$= 50.3 + 11.6 = 61.9 \text{ moles}$$

One can now find the mole fractions:

$$\text{mole fraction} = \frac{\text{number of moles of each component}}{\text{total number of moles in system}}$$

$$\text{mole fraction } H_2O = \frac{50.3}{61.9} = 0.813$$

$$\text{mole fraction NaOH} = \frac{11.6}{61.9} = 0.187.$$

9.3 The Solution Process

Solvation is the interaction of solvent molecules with solute molecules or ions to form aggregates, the particles of which are loosely bonded together.

When water is used as the solvent, the process is also called aquation or hydration.

When one substance is soluble in all proportions with another substance, then the two substances are completely miscible. Ethanol and water are a familiar pair of completely miscible substances.

A saturated solution is one in which solid solute is in equilibrium with dissolved solute.

The solubility of a solute is the concentration of dissolved solute in a saturated solution of that solute.

Unsaturated solutions contain less solute than is required for saturation.

Supersaturated solutions contain more solute than is required for saturation. Supersaturation is a metastable state; the system will revert spontaneously to a saturated solution (stable state).

9.4   Heat of Solution

Heat of solution, ΔH_{soln}, is the quantity of energy that is absorbed or released when a substance enters solution:

$$\Delta H_{soln} = \Sigma H_{products} - \Sigma H_{reactants}.$$

The magnitude of the heat of solution provides the information about the relative forces of attraction between the various particles that make up a solution.

Solutions in which the solute-solute, solute-solvent, and solvent-solvent interactions are all the same are called ideal solutions.

Problem Solving Example:

Q Using the information in the following table, what is the ΔH_{soln} for the synthesis of the gaseous solution described by the equation $CO\ (g) + Cl_2\ (g) \rightarrow COCl_2\ (g)$?

Compound	H (kcal/mole)
CO (g)	−26.4
Cl$_2$ (g)	0
COCl$_2$ (g)	−53.3

$\Delta H_{soln} = \Sigma H_{products} - \Sigma H_{reactants}$

$\Delta H_{COCl_2} = H_{COCl_2} - (H_{CO} + H_{Cl_2})$

$\Delta H_{COCl_2} = -53.3 - (-26.4 + 0) = -26.9$ kcal/mole

9.5 Solubility and Temperature

The solubility of most solids in liquids usually increases with increasing temperature.

For gases in liquids the solubility usually decreases with increasing temperature.

A positive $\Delta H°$ indicates that solubility increases with increasing temperature.

$$\log \frac{K_2}{K_1} = \frac{-\Delta H°}{2.303R}\left[\frac{1}{T_2} - \frac{1}{T_1}\right]$$

where K_2 = solubility constant at T_2, K_1 = solubility constant at T_1, and $\Delta H°$ = enthalpy change at standard conditions. For most substances, when a hot concentrated solution is cooled, the excess solid crystallizes. The overall process of dissolving the solute and crystallizing it again is known as recrystallization. It is useful in purification of the solute.

Problem Solving Example:

Q For the reaction $PbSO_4 (s) \rightarrow Pb^{2+} + SO_4^{2-}$, $\Delta H = +2,990$ cal/ mole. Will the solubility of $PbSO_4$ increase or decrease with increasing temperature? $K_{sp} = 1.8 \times 10^{-8}$ at 25 °C. Find its K_{sp} at 55 °C.

 A Whether the solubility of a salt increases or decreases with an increase in temperature can be determined by an investigation of its ΔH, the enthalpy change.

A positive $\Delta H°$ suggests that solubility increases with increasing temperature. When $\Delta H°$ is positive, the first term remains a negative value. When T increases, the first term becomes a smaller negative number. Thus, K increases. Since K measures solubility, the solubility increases. To find the K_{sp} of $PbSO_4$ at 55 °C, use the fact that

$$\log \frac{K_2}{K_1} = \frac{-\Delta H^\circ}{2.303R}\left[\frac{1}{T_2} - \frac{1}{T_1}\right],$$

where K_2 = solubility constant at the temperature in Kelvin (Celsius plus 273) of T_2; K_1 = solubility constant at a temperature of T_1. Thus, if the K_{sp} at one temperature is known, the K_{sp} at another temperature can be found given ΔH°. Let T_2 = 55 °C or 328 K and T_1 = 298 K (25 °C). Thus,

$$\log \frac{K_{328}}{K_{298}} = \frac{-\Delta H^\circ}{2.303R}\left[\frac{1}{328} - \frac{1}{298}\right]$$

$$\log \frac{K_{328}}{1.8 \times 10^{-8}} = \frac{-\left(2,990 \text{ cal/mole}\right)\left(-3.07 \times 10^{-4}/\text{K}\right)}{(2.3031)1.986 \text{ cal/mole K}}$$

$$= 0.20$$

$$K_{328} = \left(1.8 \times 10^{-8}\right)\left(10^{0.20}\right)$$

$$= \left(1.8 \times 10^{-8}\right)(1.58)$$

$$= 2.8 \times 10^{-8}$$

9.6 Effect of Pressures on Solubility

Pressure has very little effect on the solubility of liquids or solids in liquid solvents.

The solubility of gases in liquid (or solid) solvents always increases with increasing pressure.

9.7   Fractional Crystallization

The differences in solubility behavior provide the basis for a useful laboratory technique, called fractional crystallization, which is frequently used for the separation of impurities from the products of a chemical reaction.

9.8   Fractional Distillation

Fractional distillation is used to separate mixtures of volatile liquids into their components.

When a mixture is boiled, it can be separated into two parts: the distillate, which is richer than the original liquid in the more volatile component, and the residue, which is richer in the less volatile component.

9.9   Vapor Pressures of Solutions

For a solution in which a nonvolatile solute is dissolved in a solvent, the vapor pressure is due to only the vapor of the solvent above the solution. This vapor pressure is given by Raoult's law:

$$P_{solution} = X_{solvent} P^{\circ}_{solvent}.$$

Problem Solving Example:

Q   A chemist dissolves 300 g of urea in 1,000 g of water. Urea is NH_2CONH_2. Assuming the solution obeys Raoult's law, determine the vapor pressure of the solvent at 0° and 100 °C. The vapor pressure of pure water is 4.6 mm Hg and 760 mm Hg at 0 °C and 100 °C, respectively.

A To solve, you must employ Raoult's law, which states P_{H_2O} $= P°_{H_2O} X_{H_2O}$, where P_{H_2O} = the partial pressure of water (the solvent), $P°_{H_2O}$ = the vapor pressure of water when pure, and X_{H_2O} is the mole fraction of H_2O in the solution. The mole fraction is equal to

$\dfrac{N_{H_2O}}{N_{solute} + N_{H_2O}}$, where N_{H_2O} = moles of H_2O and N_{solute} = moles of solute. You are told the $P°_{H_2O}$ for H_2O at the temperatures in question.

Thus, to find the vapor pressures of the solution, you must calculate X_{H_2O} and substitute into Raoult's law.

To calculate the concentration of urea, remember, moles $= \dfrac{grams}{molecular\,weight}$. Since you have 300 g of urea and the molecular weight of urea is 60.0 g/mole, you have $= \dfrac{300\,g}{60.0\,g/mole}$ or 5.00 moles of urea. Similarly, for water (MW = 18 g/mole), you have $= \dfrac{1,000\,g}{18.0\,g/mole} = 55.6$ moles of water. Thus, the mole fraction of water

$= \dfrac{55.6\,g}{5.00 + 55.6}$ 0.917. Therefore, at 0°,

$\left(P_{H_2O} = P^0_{H_2O}X_{H_2O} = (4.6\,mm\,Hg)(0.917) = 4.2\,mm\,Hg\right)$ and at 100°,

$\left(P_{H_2O} = P^0_{H_2O}X_{H_2O} = (760\,mm\,Hg)(0.917) = 697\,mm\,Hg\right)$.

9.10 Colligative Properties of Solutions

Colligative property law: the freezing point, boiling point, and vapor pressure of a solution differ from those of the pure solvent by amounts that are directly proportional to the molal concentration of the solute.

The vapor pressure of an aqueous solution is always lowered by the addition of more solute, which causes the boiling point to be raised (boiling point elevation).

The freezing point is always lowered by addition of solute (freezing point depression). The freezing point depression, ΔT_f, equals the negative of the molal freezing point depression constant, K_f, times molality (m):

$$\Delta T_f = -K_f(m).$$

Problem Solving Example:

Q The freezing point constant of toluene is 3.33 °C per mole per 1,000 g. Calculate the freezing point of a solution prepared by dissolving 0.4 mole of solute in 500 g of toluene. The freezing point of toluene is −95.0 °C.

A The freezing point constant is defined as the number of degrees the freezing point will be lowered per 1,000 g of solvent per mole of solute present. The freezing point depression is related to this constant by the following equation:

freezing point depression = molality of solute × freezing pt. constant.

The molality is defined as the number of moles per 1,000 g of solvent. Here, one is given that 0.4 moles of solute are added to 500 g of solvent; therefore, there will be 0.8 moles in 1,000 g.

$$\frac{0.4 \text{ moles}}{500 \text{ g}} = \frac{0.8 \text{ moles}}{1,000 \text{ g}}$$

The molality of the solute is thus 0.8 m. One can now find the freezing point depression. The freezing point constant for toluene is 3.33° C/m.

freezing point depression = molality × 3.33° C/m
= 0.8 m × 3.33° C/m = 3° C/m

The freezing point of toluene is thus lowered by 3 °C.

freezing point of solution $= (- 95 \text{ °C}) - 3 \text{ °C} = - 98 \text{ °C}$

The boiling point elevation, ΔT_b, equals the molal boiling point elevation constant, K_b, times molality (m):

$$\Delta T_b = K_b(m)$$

$$\Delta T = T_{\text{solution}} - T_{\text{pure solvent}}.$$

Problem Solving Example:

Q Ethanol boils at 78.5 °C. If 10 g of sucrose $(C_{12}H_{22}O_{11})$ is dissolved in 150 g of ethanol, at what temperature will the solution boil? Assume $K_b = 1.20$ °C/m for the alcohol.

A When a nonvolatile solute, such as sucrose, is dissolved in a solvent, such as ethanol, it will raise the boiling point of the solvent. The boiling point elevation can be found by using the equation

$$\Delta T_b = K_b(m),$$

where ΔT_b is the boiling point elevation, K_b is the elevation constant, and (m) is the molality of the solution. You want to determine ΔT_b and are given K_b. Thus, you need to determine the molality of the solution. Molality is defined as the number of moles of solute per 1 kg of solvent, i.e., moles solute/1 kg solvent.

Solving for molality of sucrose: (MW of $C_{12}H_{22}O_{11} = 342.0$ g/mole).

$$\text{moles of solute} = \frac{\text{grams solute}}{\text{molecular wt. solute}}$$

$$= \frac{10 \text{ g}}{342.0 \text{ g/mole}} = 2.9 \times 10^{-2} \text{ moles}$$

You have 150 g of solvent, ethanol. But molality is per 1 kg, so you must multiply 150 g by 1 kg/1,000 g.

$$\text{molality} = \frac{0.029 \text{ mole}}{\dfrac{150}{1,000} \text{kg}} = 0.19 \text{ m}$$

Thus, the elevation of the boiling point is

$$\Delta T_b = K_b(\text{m}) = (1.20 \text{ °C/m})(0.19 \text{ m}) = 0.23 \text{ °C}.$$

Thus, the boiling point of the solution is

$$78.5 \text{ °C} + 0.23 \text{ °C} = 78.7 \text{ °C}.$$

9.11 Osmotic Pressure

Osmosis is the diffusion of a solvent through a semipermeable membrane into a more concentrated solution.

The osmotic pressure of a solution is the minimum pressure that must be applied to the solution to prevent the flow of solvent from pure solvent into the solution.

The osmotic pressure π for a solution is

$$\pi = CRT,$$

where π is the osmotic pressure, C is the concentration in molarity, R is the gas constant, and T is the temperature (K). (Note the formal similarity of the osmotic pressure equation to the ideal gas law, $C = n/V$.)

Solutions that have the same osmotic pressure are called isotonic solutions.

Reverse osmosis is a method for recovering pure solvent from a solution.

Problem Solving Example:

 A sugar solution was prepared by dissolving 9 g of sugar in 500 g of water. At 27 °C, the osmotic pressure was measured as 2.46 atm. Determine the molecular weight of the sugar.

 The molecular weight of the sugar is found by determining the concentration, C, of sugar from the equation for osmotic pressure,

$$\pi = CRT,$$

where π is the osmotic pressure, R = universal gas constant 0.082 L-atm/mole K, and T is the absolute temperature.

The osmotic pressure is measured as π = 2.46 atm and the absolute temperature is T = 27 °C + 273 = 300 K, hence

$$\pi = CRT,$$

$$2.46 \text{ atm} = C \times 0.082 \text{ L-atm/mole K} \times 300 \text{ K,}$$

or $\quad C = \dfrac{2.46 \text{ atm}}{0.082 \text{ L-atm/mole K} \times 300 \text{ K}}$

$$= 0.10 \text{ mole/L.}$$

If we assume that the volume occupied by the sugar molecules is so small it can be neglected, then in 1 L of solution there is approximately 1 L of water, or 1,000 g of water, and

$$C = 0.100 \text{ mole/1,000 g.}$$

Therefore, there is 0.100 mole of sugar dissolved in 1,000 g of water.

9 g of sugar dissolved in 500 g of water is equivalent to 18 g of sugar dissolved in 1,000 g of water (9 g/500 g = 18 g/1,000 g). But since C = 0.100 mole/1,000 g, the 18 g of sugar must correspond to 0.100 mole of sugar. Therefore, the molecular weight of the sugar is

$$18 \text{ g/0.100 mole} = 180 \text{ g/mole.}$$

9.12 Interionic Attractions

$$i = \frac{\left(\Delta T_f\right) \text{ measured}}{\left(\Delta T_f\right) \text{ calculated as nonelectrolyte}}$$

i, the van't Hoff factor, is defined as the ratio of the observed freezing point depression produced by a solute in solution to the freezing point that the solution would exhibit if the solute were a nonelectrolyte. For example, since NaCl yields 2 moles of dissolved particles (Na$^+$ and Cl$^-$) in water, its van't Hoff factor is 2 and a 1 molal solution of NaCl (aq) yields a freezing point depression which is twice as large as that produced by sucrose, a non electrolyte.

Problem Solving Example:

Q If CaCl$_2$ were a nonelectrolyte, the freezing point depression of a 0.1 m solution of CaCl$_2$ in water would be 0.186 °C. What would be the actual freezing point depression of this solution?

A For a nonelectrolyte, the van't Hoff factor, i, is equal to 1 since it does not ionize in solution. However, CaCl$_2$ does ionize in water to yield more than 1 particle per mole. In fact, it yields 1 mole of Ca^{2+} and 2 moles of Cl$^-$; therefore, its van't Hoff factor is 3 and the freezing point depression of the solution will be three times that of a nonelectrolyte. Since

$$i = \frac{\left(\Delta T_f\right) \text{ measured}}{\left(\Delta T_f\right) \text{ calculated as nonelectrolyte}},$$

ΔT_f measures = $i \times (\Delta T_f)$ calculated as nonelectrolyte

ΔT_f measures = 3×0.186 °C = 0.558 °C.

Quiz: Properties of Solutions

1. How many grams of $NaHCO_3$ must be added to water to produce 200 mL of 0.5 M solution?

    (A) 8.4 g   (D) 66.0 g

    (B) 21.0 g   (E) 84.0 g

    (C) 42.0 g

2. A saturated solution of KNO_3 contains 63 g KNO_3 at 40 °C. If a solution at the same temperature is found to contain more than 63 g of KNO_3, but with no precipitation, then the solution is probably

    (A) diluted.   (D) saturated.

    (B) concentrated.   (E) supersaturated.

    (C) unsaturated.

3. The solubility of a gas in a liquid is increased by

    (A) increased pressure.

    (B) increased temperature.

    (C) decreased pressure and increased temperature.

    (D) both (A) and (B).

    (E) None of the above.

4. The vapor pressure of pure benzene and toluene at 60 °C are 385 and 139 torr, respectively. The mole fraction of toluene in the solution is 0.60. Calculate the total vapor pressure of the solution.

    (A) 237.4 torr   (D) 61.2 torr

    (B) 83.4 torr   (E) 435.2 torr

    (C) 76.9 torr

5. All of the following are colligative properties EXCEPT

    (A) osmotic pressure.

    (B) pH of buffer solutions.

    (C) vapor pressure.

    (D) boiling point elevation.

    (E) freezing point depression.

6. What is the osmotic pressure of a 2.0 M solution of a nonelectrolyte at 298 K?

    (A) 60 atm   (D) 61 atm

    (B) 49 atm   (E) 53 atm

    (C) 30 atm

7. The most common type of solution consists of

    (A) two liquids.

    (B) two solids.

    (C) two gases.

    (D) a gas dissolved in a liquid.

    (E) a solute dissolved in a liquid.

8. The heat of solution is

    (A) the quantity of energy that is absorbed or released when a substance enters solution.

    (B) a measure of the relative forces of attraction between the various particles that make up a solution.

    (C) very large in an ideal solution.

    (D) both (A) and (B).

    (E) both (B) and (C).

9. Which of the following statements regarding solubility is true?

    (A) The solubility of most solids in liquids usually increases with increasing temperature.

    (B) The solubility of most solids in liquids usually decreases with increasing temperature.

    (C) The solubility of most gases in liquids usually increases with increasing temperature.

    (D) The solubility of most gases in liquids usually decreases with decreasing temperature.

    (E) The solubility of both liquids and gases is unchanged by changing temperature.

10. The method used to separate mixtures of volatile liquids into their components is called

    (A) fractional crystallization.

    (B) fractional distillation.

    (C) fractional solubility.

    (D) colligative separation.

    (E) aquation.

ANSWER KEY

1.	(A)	6.	(B)
2.	(E)	7.	(E)
3.	(A)	8.	(D)
4.	(A)	9.	(A)
5.	(B)	10.	(B)

Acids and Bases

10.1 Definitions of Acids and Bases

10.1.1 Arrhenius Theory

The Arrhenius theory states that acids are substances that ionize in water to give H^+ ions, and bases are substances that produce OH^- ions in water.

10.1.2 Bronsted-Lowry Theory

This theory defines acids as proton donors and bases as proton acceptors.

10.1.3 Lewis Theory

This theory defines an acid as an electron-pair acceptor and a base as an electron-pair donor.

Problem Solving Example:

Can I^+ (the iodine cation) be called a Lewis base? Explain your answer.

A Lewis base may be defined as an electron-pair donor. Writing out its electronic structure is the best way to answer this question, because it will show the existence of any available electron pairs.

The electronic structure of I^+ may be written as $\left[\; : \ddot{I} : \;\right]^+$. There are three available electron pairs. This might lead one to suspect that it is indeed a Lewis base. But note that I^+ does not have a complete octet of electrons; it does not obey the octet rule. According to this rule, atoms react to obtain an octet (8) of electrons. This confers stability.

Therefore, I^+ would certainly rather gain two more electrons than lose six. In reality, then, I^+ is an electron-pair acceptor. Such substances are called Lewis acids.

10.2 Properties of Acids and Bases

An acid is a substance that, in aqueous form, conducts electricity, has a sour taste, turns blue litmus red, reacts with active metals to form hydrogen gas, H_2, and neutralizes bases.

A base is a substance that, in aqueous form, conducts electricity, has a bitter taste, turns red litmus blue, feels soapy, and neutralizes acids.

The base that results when an acid donates a proton is called the conjugate base of the acid.

The acid that results when a base accepts a proton is called the conjugate acid of the base.

10.3 Factors Influencing the Strengths of Acids

The greater the number of oxygens bound to the element E in a hydroxy compound, H_xEO_y, the stronger is the acid. This is also a positive correlation with the oxidation state of E.

The acidity of an OH bond in M–O–H depends on the ability of M to draw electrons to itself, thereby weakening the O–H bond, making it more acidic.

With metal ions, the acidity of their solutions depends on the charge on the metal ion.

As we go down a group on the periodic table, the strengths of the oxoacids decrease.

A metal may form acidic hydroxy compounds if the metal has a high oxidation number.

CHAPTER 11

Acid-Base Equilibria in Aqueous Solutions

11.1 Ionization of Water, pH

For the equation

$$H_2O + H_2O \rightleftarrows H_3O^+ + OH^-,$$

$$K_w = \left[H_3O^+ \right]\left[OH^- \right] \left(\text{or } K_w = \left[H^+ \right]\left[OH^- \right] \right)$$

$$= 1.0 \times 10^{-14} \text{ at } 25 \text{ °C},$$

where $[H_3O^+][OH^-]$ is the product of ionic concentrations, and K_w is the ion product constant for water (or simply the ionization constant or dissociation constant).

$$pH = -\log[H^+]$$

$$pOH = -\log[OH^-]$$

Problem Solving Example:

Assuming complete ionization, calculate (a) the pH of 0.0001 N HCl and (b) the pOH of 0.0001 N KOH.

(a) pH is defined as the negative log of the hydrogen ion concentration.

$$pH = -\log[H^+]$$

The normality of an acid is defined as the number of equivalents of H^+ per liter of solution. The ionization of HCl can be written

$$HCl \rightleftharpoons H^+ + Cl^-.$$

This means that there is one H^+ for every HCl and that the concentration of H^+ equals the concentration of completely ionized HCl.

$$[H^+] = [HCl]$$

We are told that [HCl] = 0.0001 N = 1×10^{-4} N. Therefore, $[H^+] = 1 \times 10^{-4}$ N. We can now solve for pH. Note: In this problem, normality = molarity (concentration), since equivalent weight = MW.

$$pH = -\log[H^+]$$

$$pH = -\log(1 \times 10^{-4}) = 4$$

The pH of this solution is 4.

(b) The pOH is defined as the negative log of the OH^- ion concentration.

$$pOH = -\log[OH^-]$$

The ionization of KOH can be stated

$$KOH \overset{\rightarrow}{\underset{\leftarrow}{}} K^+ + OH^-.$$

Therefore, one OH is formed for every KOH, and when KOH is completely ionized, their concentrations are equal.

$$[KOH] = [OH^-]$$

We are told that $[KOH] = 0.0001$ N; thus $[OH^-] = 0.0001$ N (again, normality = molarity).

Solving for pOH:

$$pOH = -\log[OH^-]$$

$$= -\log(0.0001) = -\log(1 \times 10^{-4}) = 4.$$

The pOH of this solution is 4.

Note that the sum of pH and pOH is:

$$pK_w = pH + pOH = 14.0.$$

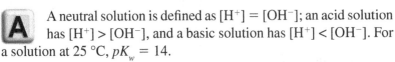

Problem Solving Example:

What is the pH of a neutral solution at 50 °C where $pK_w = 13.26$ at 50 °C?

A neutral solution is defined as $[H^+] = [OH^-]$; an acid solution has $[H^+] > [OH^-]$, and a basic solution has $[H^+] < [OH^-]$. For a solution at 25 °C, $pK_w = 14$.

pK_w indicates the amount of dissociation of water. To find the neutral pH, one lets pH = pOH = x. Since

$$pH + pOH = 2x = 14 = pK_w,$$

$$x = 7.$$

However, the solution in question is at 50 °C. At 50 °C, $pK_w = 13.26$. Therefore, to find the neutral pH,

$$pH + pOH = 2x = pK_w = 13.26$$

$$x = 6.63 = \text{neutral pH.}$$

In a neutral solution, pH = 7.0. In an acidic solution, pH is less than 7.0. In basic solutions, pH is greater than 7.0. The smaller the pH, the more acidic is the solution. Note that since pK_w (like all equilibrium constants) varies with temperature, neutral pH is less than (or greater than) 7.0 when the temperature is higher than (or lower than) 25 °C.

11.2 Dissociation of Weak Electrolytes

For the equation

$$A^- + H_2O \rightleftharpoons HA + OH^-,$$

$$K_b = \frac{[HA][OH^-]}{[A^-]}$$

where K_b is the base ionization constant.

$$K_a = \frac{[H^+][A^-]}{[HA]}$$

where K_a is the acid ionization constant.

$$K_b = \frac{K_w}{K_a}$$

for any conjugate acid-base pair, and therefore, $K_w = [H^+][OH^-]$.

Problem Solving Example:

 The ionization constant for acetic acid is 1.8×10^{-5}. Calculate the concentration of H^+ ions in a 0.10 molar solution of acetic acid.

The ionization constant (K_a) is defined as the concentration of H^+ ions times the concentration of the conjugate base ions of a given acid divided by the concentration of unionized acid. For an acid, HA,

$$K_a = \frac{\left[H^+\right]\left[A^-\right]}{[HA]},$$

where K_a is the ionization constant, $[H^+]$ is the concentration of H^+ ions, $[A^-]$ is the concentration of the conjugate base ions, and $[HA]$ is the concentration of unionized acid. The K_a for acetic acid is stated as

$$K_a = \frac{\left[H^+\right]\left[\text{acetate ion}\right]}{\left[\text{acetic acid}\right]} = 1.8 \times 10^{-5}.$$

The chemical formula for acetic acid is $HC_2H_3O_2$. When it is ionized, one H^+ is formed and one $C_2H_3O_2^-$ (acetate) is formed, thus the concentration of H^+ equals the concentration of $C_2H_3O_2^-$.

$$[H^+] = [C_2H_3O_2^-]$$

The concentration of unionized acid is decreased when ionization occurs. The new concentration is equal to the concentration of H^+ subtracted from the concentration of unionized acid.

$$[HC_2H_3O_2] = 0.10 - [H^+]$$

Since $[H^+]$ is small relative to 0.10, one may assume that $0.10 - [H^+]$ is approximately equal to 0.10.

$$0.10 - [H^+] \cong 0.10$$

Using this assumption, and the fact that $[H^+] = [C_2H_3O_2^-]$, K_a can be rewritten as

$$K_a = \frac{[H^+][H^+]}{[0.10]} = 1.8 \times 10^{-5}.$$

Solving for the concentration of H+ :

$$[H^+]^2 = (1.0 \times 10^{-1})(1.8 \times 10^{-5}) = 1.8 \times 10^{-6}$$

$$[H^+] = \sqrt{1.8 \times 10^{-6}} = 1.3 \times 10^{-3}$$

The concentration of H^+ is thus 1.3×10^{-3} M.

11.3 Dissociation of Polyprotic Acids

For $H_2S \rightleftharpoons H^+ + HS^-$, $K_{a_1} = \dfrac{[H^+][HS^-]}{[H_2S]}$.

For $HS^- \rightleftharpoons H^+ + S^{2-}$, $K_{a_2} = \dfrac{[H^+][S^{2-}]}{[HS^-]}$.

K_{a_1} is much greater than K_{a_2}.

Also, $K_a = K_{a_1} \times K_{a_2} = \dfrac{[H^+][HS^-]}{[H_2S]} \times \dfrac{[H^+][S^{2-}]}{[HS^-]}$

$$= \frac{[H^+]^2[S^{2-}]}{[H_2S]}.$$

This last equation is useful only in situations where two of the three concentrations are given and you wish to calculate the third.

Problem Solving Example:

Q Write the equations for the stepwise dissociation of pyrophosphoric acid, $H_4P_2O_7$. Identify all conjugate acid-base pairs.

A Pyrophosphoric acid is an example of a polyprotic acid. Polyprotic acids furnish more than one proton per molecule. From its molecular formula, $H_4P_2O_7$, one can see there exist four hydrogen atoms. This might lead one to suspect that it is tetraprotic, i.e., having four protons that can be donated per molecule. This is in fact the case, which means there exist four dissociation reactions. In general, the equation for a dissociation reaction is

$$HA + H_2O \rightarrow H_3O^+ + A^-.$$

Polyprotic acids follow this pattern. Thus, one can write the following equations for the stepwise dissociation of $H_4P_2O_7$.

(1) $H_4P_2O_7 + H_2O \rightarrow H_3O^+ + H_3P_2O_7^-$

(2) $H_3P_2O_7^- + H_2O \rightarrow H_3O^+ + H_2P_2O_7^{2-}$

(3) $H_2P_2O_7^{2-} + H_2O \rightarrow H_3O^+ + HP_2O_7^{3-}$

(4) $HP_2O_7^{3-} + H_2O \rightarrow H_3O^+ + P_2O_7^{4-}$

To identify all conjugate acid-base pairs, note the definition of the term. The base that results when an acid donates its proton is called the conjugate base. The acid that results when a base accepts a proton is called the conjugate acid. From these definitions, one sees that in all cases H_3O^+ is the conjugate acid of H_2O (the base in these reactions) and $H_3P_2O_7^-$, $H_2P_2O_7^{2-}$, $HP_2O_7^{3-}$, and $P_2O_7^{4-}$ are the conjugate bases of $H_4P_2O_7$, $H_3P_2O_7^-$, $H_2P_2O_7^{2-}$, and $HP_2O_7^{3-}$, respectively.

11.4 Buffers

Buffer solutions are equilibrium systems that resist changes in acidity and maintain constant pH when acids or bases are added to them.

The most effective pH range for any buffer is at or near the pH where the acid and salt concentrations are equal (that is, pK_a).

The pH for a buffer is given by

$$pH = pK_a + \log \frac{\left[A^- \right]}{\left[HA \right]} = pK_a + \log \frac{\left[base \right]}{\left[acid \right]},$$

which is obtained very simply from the equation for weak acid equilibrium,

$$K_a = \frac{\left[H^+ \right]\left[A^- \right]}{\left[HA \right]}.$$

Problem Solving Example:

 What is the hydrogen ion concentration of a buffer solution that is 0.05 M in acetic acid and 0.1 M in sodium acetate? (The K_a for acetic acid is 1.8×10^{-5}.)

 Acetic acid ionizes,

$$HC_2H_3O_2 \rightleftarrows H^+ + C_2H_3O_2^-,$$

and sodium acetate completely dissociates,

$$NaC_2H_3O_2 \rightarrow Na^+ + C_2H_3O_2^-.$$

The solution is a buffer solution since it is composed of a weak acid and a salt of the weak acid. The hydrogen ion concentration of such a buffer can be calculated from the following expression.

Since $K_a = \dfrac{\left[H^+\right]\left[A^-\right]}{[HA]}$ where $[A^-]$ = concentration of the salt of acid and $[HA]$ = concentration of the acid,

$$\left[H^+\right] = \frac{[\text{acid}]}{\text{salt}} \times K_a$$

$$\left[H^+\right] = \frac{0.05}{0.1} \times \left(1.8 \times 10^{-5}\right)$$

$$= 0.9 \times 10^{-5}, \text{ or } 9 \times 10^{-6} \text{ M.}$$

11.5 Hydrolysis

Hydrolysis refers to the action of salts of weak acids or bases with water to form acidic or basic solutions.

Salts of Weak Acids and Strong Bases: Anion Hydrolysis

For $C_2H_3O_2^- + H_2O \rightleftarrows HC_2H_3O_2 + OH^-$,

$$K_h = \frac{\left[HC_2H_3O_2\right]\left[OH^-\right]}{\left[C_2H_3O_2^-\right]}$$

where K_h is the hydrolysis constant for the acetate ion, which is just K_b for acetate.

Also,

$$K_a = \frac{\left[H^+\right]\left[C_2H_3O_2^-\right]}{\left[HC_2H_3O_2\right]}$$

and

$$K_h = \frac{K_w}{K_a}.$$

Salts of Strong Acids and Weak Bases: Cation Hydrolysis

For $NH_4^+ + H_2O \rightleftarrows H_3O^+ + NH_3$,

$$K_h = \frac{\left[H_3O^+\right]\left[NH_3\right]}{\left[NH_4^+\right]}$$

$$\left(K_h = K_a \text{ for } NH_4^+\right);$$

also,

$$K_h = \frac{K_w}{K_b}$$

and

$$K_b = \frac{\left[NH_4^+\right]\left[OH^-\right]}{\left[NH_3\right]}.$$

Hydrolysis of Salts of Polyprotic Acids

For $S^{2-} + H_2O \rightleftarrows HS^- + OH,$

$$K_{h_1} = \frac{K_w}{K_{a_2}} = \frac{\left[HS^-\right]\left[OH^-\right]}{\left[HS^{2-}\right]},$$

where K_{a_2} is the acid dissociation constant for the weak acid HS^-.

For $HS^- + H_2O \rightleftarrows H_2S + OH^-,$

$$K_{h_2} = \frac{K_w}{K_{a_1}} = \frac{[H_2S][OH^-]}{[HS^-]},$$

where K_{a_1} is the dissociation constant for the weak acid H_2S.

Problem Solving Example:

If the hydrolysis constant of Al^{3+} is 1.4×10^{25}, what is the concentration of H_3O^+ in 0.1 M $AlCl_3$?

Hydrolysis refers to the action of the salts of weak acids and bases with water to form acidic or basic solutions. Consequently, to answer this question write out the reaction, which illustrates this hydrolysis, and write out an equilibrium constant expression. From this, the concentration of H_3O^+ can be defined. The net hydrolysis reaction is

$$AlCl_3 \rightarrow Al^{3+} + 3Cl^-$$

$$Al^{3+} + H_2O \rightleftharpoons Al(OH)^{2+} + H_3O^+$$

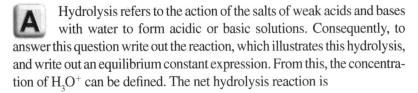

$$K_h = 1.4 \times 10^{-5} = \frac{[H_3O^+][Al(OH)^{2+}]}{[Al^{3+}]}.$$

Water is excluded in this expression since it is considered as a constant. Let x = the moles/L of $[H_3O^+]$. Since H_3O^+ and $Al(OH)^{2+}$ are formed in equal mole amounts, the concentration of $[Al(OH)^{2+}]$ can also be represented by x. If one starts with 0.1 M of Al^{3+}, and x moles/L of it forms H_3O^+ [and $Al(OH)^{2+}$], one is left with $0.1 - x$ at equilibrium. Substituting these representations into the K_h expression,

$$\frac{x \times x}{0.1 - x} = 1.4 \times 10^{-5}.$$

If one solves for x, the answer is $x = 1.2 \times 10^{-3}$ M, which equals $[H_3O^+]$.

11.6 Acid-Base Titration: The Equivalence Point

Titration is the process of determining the amount of a solution of known concentration that is required to react completely with a certain amount of a sample that is being analyzed.

The solution of known concentration is called a standard solution, and the sample being analyzed is the unknown.

$$V_A \times N_A = \text{equiv } a, \text{ and } V_B \times N_B = \text{equiv } b,$$

where V is the volume in liters and N is the normality. An equivalent is defined as the weight in grams of a substance that releases 1 mole of either protons (H^+) or hydroxyl ions (OH^-).

$$V_A \times N_A = V_B \times N_B$$

at the equivalence point, since $eq_a = eq_b$.

An endpoint is the point at which a particular indicator changes color.

The equivalence point occurs when equal numbers of equivalents of acid and base have been reacted.

11.6.1 Strong Acid-Strong Base Titration

At the equivalence point, the solution is neutral because neither of the ions of the salt in solution undergoes hydrolysis.

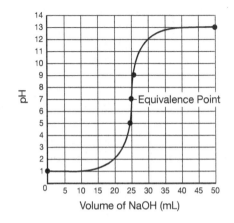

Figure 11.1–Titration of 0.1 M HCl with 0.1 M NaOH.

Problem Solving Example:

Q A 50.0 mL solution of sulfuric acid was found to contain 0.49 g of H_2SO_4. This solution was titrated against a sodium hydroxide solution of unknown concentration. 12.5 mL of the acid solution was required to neutralize 20.0 mL of the base. What is the concentration of the sodium hydroxide solution?

A At the neutralization point, the number of equivalents of acid in the 12.5 mL volume is equal to the number of equivalents of base in the 20.0 mL volume. Since the normality is defined as the number of equivalents per liter of solution, the number of equivalents is equal to the normality times the volume. At the neutralization point we have

$$N_a V_a = N_b V_b$$

where N_a = normality of acid, V_a = volume of acid, N_b = normality of base, and V_b = volume of base.

The normality of the 50.0 mL (0.0500 L) sulfuric acid solution is

$$N_a = \frac{\text{number of eq. in } 0.0500\,L}{0.0500\,L}$$

$$= \frac{\text{mass of acid / gram eq. weight}}{0.0500\,L}.$$

The gram equivalence weight of sulfuric acid is 49.1 g/eq. because there are 2 equivalents per molecule. The MW of H_2SO_4 is 98.1 g/mole.

$$N_a = \frac{\text{mass of acid / gram eq. weight}}{0.0500\,L}$$

$$= \frac{0.49\,g / 49.1\,g / eq.}{0.0500\,L} = 0.20 \text{ eq./L}$$

$$= 0.20 \text{ N}$$

The normality of the base is then found as follows:

$$N_a V_a = N_b V_b$$

$$N_b = \frac{N_a V_a}{V_b} = \frac{0.20\,N \times 12.5\,mL}{20.0\,mL} = 0.125\,N.$$

Therefore, the sodium hydroxide solution is 0.125 N, which because there is 1 ionizable OH^- in NaOH, is equal to 0.125 M.

11.7 Acid-Base Indicators

Indicators are used to detect the equivalence point in an acid-base titration. They are usually weak organic acids or bases that change color over a narrow pH range ($\Delta pH \approx 2$ pH units).

The choice of a suitable indicator for a particular titration depends on the pH at the equivalence point. The color change of the indicator should occur very near the pH at the equivalence point.

Quiz: Acids and Bases—Acid-Base Equilibria in Aqueous Solutions

1. Arrhenius would define a base as

    (1) something which yields hydroxide ions in solution.

    (2) a proton acceptor.

    (3) an electron pair donator.

    (A) 1, 2, and 3.   (D) 1 only.

    (B) 1 and 3.   (E) 1 and 2.

    (C) 2 only.

2. $HA + B \rightarrow H^+ + B + A^-$

    By the Bronsted theory, the acid in this equation is

    (A) A^-.   (D) B.

    (B) HA.   (E) HB.

    (C) AB.

3. Which of the following can function as a Lewis acid?

    (A) $:\overset{..}{\underset{..}{I}}\!:^+$ (D) $:NH_3$

    (B) $:CN^-$ (E) $:\overset{..}{\underset{..}{Br}}\!:^-$

    (C) $CH_3 - \overset{..}{\underset{..}{O}} - CH_3$

4. Which of the following is most acidic?

    (A) $HClO_4$ (D) HCN

    (B) HF   (E) HCl

    (C) H_3PO_4

5. Which of the following indicates a basic solution?

    (A) $[H^+] > 10^{-7}$ (D) pH = 7

    (B) $[OH^-] < 10^{-10}$ (E) pH = 9

    (C) pH = 5

6. What is the concentration of $[OH^-]$ in a NH_4OH solution if it has a pH of 11?

    (A) 10^{-1} (D) 10^{-9}

    (B) 10^{-3} (E) 10^{-11}

    (C) 10^{-5}

7. A buffer solution was prepared by mixing 100 mL of a 1.2 M NH_3 solution and 400 mL of a 0.5 M NH_4Cl solution. What is the pH of this buffer solution, assuming a final volume of 500 mL and $K_b = 1.8 \times 10^{-5}$?

    (A) 1.08   (D) 9

    (B) 4.96   (E) 8

    (C) 5.8

8. Hydrolysis of sodium carbonate yields

    (A) a strong acid and a strong base.

    (B) a weak acid and a strong base.

    (C) a weak acid and a weak base.

    (D) a strong acid and a weak base.

    (E) None of the above.

9. How many mL of 5 M NaOH are required to completely neutralize 2 L of 3 M HCl?

   (A) 600 mL   (D) 1,500 mL

   (B) 900 mL   (E) 1,800 mL

   (C) 1,200 mL

10. A base is a substance that

    (A) does not conduct electricity.

    (B) turns litmus paper from blue to red.

    (C) has a sour taste.

    (D) feels soapy.

    (E) reacts with active metals to form hydrogen.

ANSWER KEY

1.	(D)	6.	(B)
2.	(B)	7.	(D)
3.	(A)	8.	(E)
4.	(A)	9.	(C)
5.	(E)	10.	(D)

CHAPTER 12

Chemical Equilibrium

12.1 The Law of Mass Action

At equilibrium, both the forward and reverse reactions take place at the same rate, and thus the concentrations of reactants and products no longer change with time.

The law of mass action states that the rate of an elementary chemical reaction is proportional to the product of the concentrations of the reacting substances, each raised to its respective stoichiometric coefficient.

For the reaction $aA + bB \rightleftarrows eE + fF$, at constant temperature,

$$K_c = \frac{[E]^e [F]^f}{[A]^a [B]^b},$$

where K_c is the equilibrium constant.

The entire relationship is known as the law of mass action.

$\dfrac{[E]^e [F]^f}{[A]^a [B]^b}$ is known as the mass action expression. Note that if any of the species (A, B, E, F) is a pure solid or pure liquid, it does not appear in the expression for K_c.

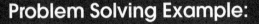

Problem Solving Example:

The following reaction,

$$2H_2S(g) \rightleftarrows 2H_2(g) + S_2(g),$$

was allowed to proceed to equilibrium. The contents of the 2 L reaction vessel were then subjected to analysis and found to contain 1.0 mole H_2S, 0.20 mole H_2, and 0.80 mole S_2. What is the equilibrium constant K for this reaction?

This problem involves substitution into the equilibrium constant expression for this reaction,

$$K = \frac{[H_2]^2 [S_2]}{[H_2S]^2}.$$

The equilibrium concentration of the reactant and products are $[H_2S]$ = 1.0 mole/2 L = 0.50 M, $[H_2]$ = 0.20 mole/2 L = 0.10 M, and $[S_2]$ = 0.80 mole/2 L = 0.40 M. Hence, the value of the equilibrium constant is

$$K = \frac{[H_2]^2 [S_2]}{[H_2S]^2} = \frac{(0.10)^2 (0.40)}{(0.50)^2} = 0.016$$

for this reaction.

For the reaction

$$N_2(g) + 3H_2(g) \rightleftarrows 2NH_3(g),$$

$$K_p = \frac{\left(P_{NH_3}\right)^2}{P_{N_2}\left(P_{H_2}\right)^3}$$

where K_p is the equilibrium constant derived from partial pressures.

Problem Solving Example:

 At 395 K, chlorine and carbon monoxide react to form phosgene, $COCl_2$, according to the equation

$$CO(g) + Cl_2(g) \rightleftharpoons COCl_2(g).$$

The equilibrium partial pressures are $P_{Cl_2} = 0.128$ atm, $P_{CO} = 0.116$ atm, and $P_{COCl_2} = 0.334$ atm. Determine the equilibrium constant K_p for the dissociation of phosgene.

A After obtaining an expression for K_p for the dissociation of phosgene, the degree of dissociation under 1 atm of total pressure will be obtained by combining K_p with Dalton's law of partial pressures.

The dissociation of phosgene may be written as

$$COCl_2(g) \rightleftharpoons CO(g) + Cl_2(g).$$

By definition, K_p is the product of the partial pressures of the products divided by the product of the partial pressure of the reactants. Hence,

$$K_p = \frac{P_{CO}P_{Cl_2}}{P_{COCl_2}} = \frac{(0.116\,\text{atm})(0.128\,\text{atm})}{(0.334\,\text{atm})} = 0.0445\,\text{atm}.$$

K_p for the dissociation of phosgene is thus 0.0445 atm.

12.2 Kinetics and Equilibrium

The rate of an elementary chemical reaction is proportional to the concentrations of the reactants raised to powers equal to their coefficients in the balanced equation.

For $aA + bB \rightleftarrows eE + fF$,

  $rate_f = k_f[A]^a[B]^b$,

  $rate_r = k_r[E]^e[F]^f$,

and

$$\frac{k_f}{k_r} = \frac{[E]^e\,[F]^f}{[A]^a\,[B]^b} = K_c,$$

where k_f and k_r are rate constants for the forward and reverse reactions, respectively.

Problem Solving Example:

For the reaction $H_2(g) + I_2(g) \rightleftarrows 2HI(g)$, when $[H_2] = [I_2] = 0.92\,M$ and $[HI] = 0.34$ M. What is the rate of the forward reaction, k_f, the reverse reaction, k_r, and the equilibrium constant, K_c?

A
$$K_c = \frac{k_f}{k_r} = \frac{[H]^2}{[H_2][I_2]}$$

  Therefore, $k_f = [HI]^2 = (0.34\,M)^2 = 0.12\,M^2$

and $k_r = [H_2][I_2] = (0.92\,M)(0.92\,M) = 0.85\,M^2$

        $K_c = k_f/k_r = 0.12\,M^2/0.85\,M^2 = 0.14.$

12.3 Thermodynamics and Chemical Equilibrium

$$\Delta G = \Delta G° + 2.303RT \log Q$$

The symbol Q represents the mass action expression for the reaction. For gases, Q is written with partial pressures. ΔG is the free energy.

At equilibrium $Q = K_{eq}$, and the products and reactants have the same total free energy, such that $\Delta G = 0$.

$$\Delta G° = -2.303RT \log K_{eq} = -RT \ln K_{eq}$$

For the equation $2NO_2(g) \rightleftarrows 2N_2O_4(g)$,

$$\Delta G° = -2.303\,RT \log \left(\frac{P_{N_2O_4}}{\left(P_{NO_2}\right)^2} \right)_{eq}, K_c = \frac{[N_2O_4]}{[NO_2]^2}.$$

Problem Solving Example:

 In the human body, the enzyme phosphoglucomutase catalyzes the conversion of glucose-1-phosphate into glucose-6-phosphate:

glucose-1-phosphate $\rightleftarrows$ glucose-6-phosphate.

At 38 °C, the equilibrium constant, K, for this reaction is approximately 20. Calculate the free energy change, $\Delta G°$, for the equilibrium conversion. Calculate the free energy change ΔG for the nonequilibrium situation in which [glucose-1-phosphate] = 0.001 M and [glucose-6-phosphate] = 0.050 M.

 The nonequilibrium free energy change ΔG is related to the standard free energy change $\Delta G°$ by

$$\Delta G = \Delta G° + 2.303RT \log Q,$$

where R is the gas constant and T the absolute temperature.

At equilibrium, $\Delta G = 0$, hence

$$\Delta G° = -2.303RT \log K.$$

Thus, the equilibrium conversion with $K = 20$ and $T = 38°$ C = 311 K.

$$\Delta G° = -2.303 RT \log K$$

$$= -2.303 \times 1.986 \text{ cal/mole K} \times 311 \text{ K} \times \log 20$$

$$= -1,900 \text{ cal/mole}$$

The equilibrium constant for the conversion glucose-1-phosphate $\rightleftarrows$ glucose-6-phosphate is

$$K = \frac{[\text{glucose-6-phosphate}]}{[\text{glucose-1-phosphate}]}.$$

In the case where [glucose-6-phosphate] = 0.050 M and [glucose-1-phosphate] = 0.001 M,

$$Q = \frac{[\text{glucose-6-phosphate}]}{[\text{glucose-1-phosphate}]} = \frac{0.050 \text{ M}}{0.001 \text{ M}} = 50,$$

where the Q has been used to distinguish this ratio from the equilibrium constant. Then

$$\Delta G = \Delta G° + 2.303 RT \log Q$$

$$= -1,900 \text{ cal/mole} + (2.303)(1.986 \text{ cal/mole K})(311 \text{ K})(\log 50)$$

$$= -1,900 \text{ cal/mole} + 2,400 \text{ cal/mole}$$

$$= +500 \text{ cal/mole}.$$

12.4 The Relationship Between K$_p$ and K$_c$

$$K_p = \frac{P_E^e P_F^f}{P_A^a P_B^b} = \frac{[\text{E}]^e (RT)^e [\text{F}]^f (RT)^f}{[\text{A}]^a (RT)^a [\text{B}]^b (RT)^b}$$

and

$$K_p = \frac{[E]^e [F]^f}{[A]^a [B]^b}(RT)^{(e+f)-(a+b)}$$

Therefore,

$$K_p = K_c(RT)^{\Delta ng}$$

where Δng is the change in the number of moles of gas upon going from reactants to products.

Problem Solving Example:

The equilibrium constant, K_c, is 4.26×10^{-3} M for the following reaction at 25 °C and 1 atm pressure: N_2O_4 (g) $\rightleftarrows$ $2NO_2$ (g). What is K_p for this reaction?

$$K_p = K_c(RT)^{\Delta ng}$$

In this equation $\Delta ng = +1$ since 1 mole of reactants is used to generate 2 moles of products. Therefore,

$$K_p = (4.26 \times 10^{-3} \text{ M})(0.082 \text{ L-atm/mole K})(298 \text{ K})$$

$$K_p = 0.104 \text{ atm.}$$

12.5 Heterogeneous Equilibria

For heterogeneous reactions, the equilibrium constant expression does not include the concentrations of pure solids and liquids.

For the equation

$$2NaHCO_3(s) \rightleftarrows Na_2CO_3(s) + CO_2(g) + H_2O(g),$$

$$K_p = P_{CO_2} P_{H_2O},$$

$$K_c = [CO_2][H_2O]$$

and $K_p = K_c(RT)^{\Delta ng}$, where $\Delta ng = +2$ for the reaction.

12.6 Le Chatelier's Principle and Chemical Equilibrium

Le Chatelier's principle states that when a system at equilibrium is disturbed by the application of a stress (change in temperature, pressure, or concentration), it reacts to minimize the stress and attain a new equilibrium position.

12.6.1 Effect of Changing the Concentrations on Equilibrium

When a system at equilibrium is disturbed by adding or removing one of the substances, all the concentrations will change until a new equilibrium point is reached with the same value of K_{eq}.

An increase in the concentrations of reactants shifts the equilibrium to the right, thus increasing the amount of products formed. Decreasing the concentrations of reactants shifts the equilibrium to the left and thus decreases the concentrations of products formed.

12.6.2 Effect of Temperature on Equilibrium

An increase in temperature causes the position of equilibrium of an exothermic reaction to be shifted to the left, while that of an endothermic reaction is shifted to the right.

12.6.3 Effect of Pressure on Equilibrium

Increasing the pressure on a system at equilibrium will cause a shift in the position of equilibrium in the direction of the fewest number of moles of gaseous reactants or products.

12.6.4 Effect of a Catalyst on the Position of Equilibrium

A catalyst lowers the activation energy barrier that must be overcome in order for the reaction to proceed. A catalyst merely speeds the approach to equilibrium but does not change K_{eq} (or $\Delta G°$) at all.

12.6.5 Addition of an Inert Gas

If an inert gas is introduced into a reaction vessel containing other gases at equilibrium, it will cause an increase in the total pressure within the container. However, this kind of pressure increase will not affect the position of equilibrium.

Problem Solving Example:

Q You are given a box in which $PCl_5(g)$, $PCl_3(g)$, and $Cl_2(g)$ are in equilibrium with each other at 546 K. Assuming that the decomposition of PCl_5 to PCl_3 and Cl_2 is endothermic, what effect would there be on the concentration of PCl_5 in the box if each of the following changes were made? (a) Add Cl_2 to the box, (b) reduce the volume of the box, and (c) raise the temperature of the system.

A You are told that the following equilibrium exists in the box, $PCl_5 \rightleftarrows PCl_3 + Cl_2$ (all gases), and asked to see what happens to $[PCl_5]$ when certain changes are made. This necessitates the use of Le Chatelier's principle, which states that if a stress is applied to a system at equilibrium, then the system readjusts to reduce the stress. With this in mind, proceed as follows:

(a) Here, you are adding Cl_2 to the box. This results in a stress, since one of the components in the equilibrium has its concentration increased. According to Le Chatelier's principle, the system will act to relieve this increased concentration of Cl_2 – the stress. It can do so if the Cl_2 combines with PCl_3 to produce more PCl_5. In this fashion, the stress is reduced but the concentration of PCl_5 is increased.

(b) When the volume of the box is reduced, the concentration of the species is increased, i.e., the molecules are crowded closer together. Thus, a stress is applied. The stress can only be relieved (Le Chatelier's principle) if the molecules could be reduced in number. Notice that in our equilibrium expression there are two molecules, one each of PCl_3 and Cl_2, are produced from one molecule of PCl_5. In other words, the number of molecules is reduced if the equilibrium shifts to the left, so that the concentration of PCl_5 increases. This is exactly what happens. As such, the $[PCl_5]$ increases.

(c) You are told that the decomposition of PCl_5 is endothermic (absorbing heat). In other words, it must absorb heat from the surroundings to proceed. If you increase the temperature, more heat is available, and the decomposition proceeds more readily, which means $[PCl_5]$ decreases. This fact can also be seen from the equilibrium constant of the reaction, K. This constant measures the ratio of products to reactants, each raised to the power of its coefficients in the chemical reaction. Now, when a reaction is endothermic, K is increased. For K to increase, the reactant's concentration must decrease. Again, therefore, you see that $[PCl_5]$ decreases.

Quiz: Chemical Equilibrium

1. What is the equilibrium constant for the following reaction?
   $aA + bB + cC \rightleftarrows dD + eE$

   (A) $[A]^a[B]^b[C]^c[D]^d[E]^e$

   (B) $[D]^d[E]^e$

   (C) $[A]^a[B]^b[C]^c$

   (D) $\dfrac{1}{[A]^a[B]^b[C]^c}$

   (E) $\dfrac{[D]^d[E]^e}{[A]^a[B]^b[C]^c}$

2. Consider the following reversible reaction:

   $$H_2 + 3N_2 \rightleftarrows 2NH_3$$

   Its equilibrium constant K is expressed as

   (A) $\dfrac{[NH_3]}{[N_2][H_2]^3}$.   (D) $[NH_3]$.

   (B) $\dfrac{[NH_3]^2}{[N_2]^3[H_2]}$.   (E) $[N_2]^2[H_2]^3$.

   (C) $\dfrac{[NH_3]}{[N_2][H_2]^3}$

3. A reaction mixture consists of N_2, H_2, and NH_3. At 298 K, what is ΔG for the following reaction? (P = pressure)
   Reaction: $N_2(g) + 3H_2(g) \rightarrow 2NH_3(g)$

   (A) $\Delta G = \Delta G° + 2.3RT \log \dfrac{P_{N_2} P_{H_2}^3}{P_{NH_3}^2}$

   (B) $\Delta G = 2.3RT \log \dfrac{P_{NH_3}^2}{P_{N_2} P_{H_2}^3}$

   (C) $\Delta G = \Delta G° + 2.3RT \log \dfrac{P_{NH_3}^2}{P_{N_2} P_{H_2}^3}$

   (D) $\Delta G = \dfrac{\Delta G° + 2.3RT}{\dfrac{P_{NH_3}^2}{P_{H_2}^3 P_{N_2}}}$

   (E) $\Delta G = \Delta G° + 2.3RT \dfrac{P_{NH_3}^2}{P_{N_2} P_{H_2}^2}$

4. For which one of the following equilibrium equations will K_p equal K_c?

   (A) $PCl_5 \rightleftarrows PCl_3 + Cl_2$

   (B) $COCl_2 \rightleftarrows CO + Cl_2$

   (C) $H_2 + I_2 \rightleftarrows 2HI$

   (D) $3H_2 + N_2 \rightleftarrows 2NH_3$

   (E) $2SO_3 \rightleftarrows 2SO_2 + O$

5. The equilibrium expression, $K_{eq} = [CO_2]$, represents the reaction

   (A) $C(s) + O_2(g) \rightleftarrows CO_2(g)$.

   (B) $CO(g) + \dfrac{1}{2}O_2(g) \rightleftarrows CO_2(g)$.

   (C) $CaCO_3(s) \rightleftarrows CaO(s) + CO_2(g)$.

   (D) $CO_2(g) \rightleftarrows C(s) + O_2(g)$.

   (E) $CaO(s) + CO_2(g) \rightleftarrows CaCO_3(s)$.

6. $N_2 + 3H_2 \rightleftarrows 2NH_3 \uparrow + heat$

   In this reversible reaction, the equilibrium shifts to the right be-cause of all the following factors EXCEPT

   (A) adding heat.

   (B) adding reactant amounts.

   (C) formation of ammonia gas.

   (D) increasing pressure on reactants.

   (E) yielding an escaping gas.

7. Which of the following shifts the equilibrium of the following reaction to the right?

   $A(g) + B(g) + C(g) \rightleftarrows A(g) + BC(g)$

   (A) Addition of more A

   (B) Removal of B

   (C) Increasing the pressure

   (D) Decreasing the temperature

   (E) Increasing the temperature

8. $\Delta H°_{298} = -46.19\ kJ \times mol^{-1}$ for the reaction below. Which state-ment is true about the equilibrium constant K_{eq} for this reaction?

   $N_2(g) + 3H_2(g) \rightleftarrows 2NH_3(g)$

   (A) K_{eq} increases with increasing temperature.

   (B) K_{eq} decreases with increasing temperature.

   (C) K_{eq} increases with increasing pressure.

   (D) K_{eq} decreases with increasing pressure.

   (E) K_{eq} is independent of temperature and pressure.

9. An increase in pressure will change the equilibrium constant by

    (A) shifting to the side where a smaller volume results.

    (B) shifting to the side where a larger volume results.

    (C) favoring the exothermic reaction.

    (D) favoring the endothermic reaction.

    (E) favoring the side with the fewest moles of material.

10. A catalyst will increase the rate of a chemical reaction by

    (A) shifting the equilibrium to the right.

    (B) lowering the activation energy.

    (C) shifting the equilibrium to the left.

    (D) increasing the activation energy.

    (E) favoring the side with the fewest moles of material.

ANSWER KEY

1.	(E)	6.	(A)
2.	(B)	7.	(C)
3.	(C)	8.	(B)
4.	(C)	9.	(E)
5.	(C)	10.	(B)

CHAPTER 13

Chemical

Thermodynamics

13.1 Some Commonly Used Terms in Thermodynamics

A system is that particular portion of the universe on which we wish to focus our attention.

Everything else is called the surroundings.

An adiabatic process occurs when the system is thermally isolated so that no heat enters or leaves.

An isothermal process occurs when the system is maintained at the same temperature throughout an experiment ($t_{final} = t_{initial}$).

An isopiestic (isobaric) process occurs when the system is maintained at constant pressure (i.e., $P_{final} = P_{initial}$).

The state of the system is some particular set of conditions of pressure, temperature, number of moles of each component, and their physical form (i.e., gas, liquid, solid, or crystalline form).

State functions depend only on the present state of the substance and not on the path by which the state was attained. Enthalpy, energy, Gibbs free energy, and entropy are examples of state functions.

Heat capacity is the amount of heat energy required to raise the temperature of a given quantity of a substance 1 degree Celsius.

Specific heat is the amount of heat energy required to raise the temperature of 1g of a substance by 1 °C.

Molar heat capacity is the heat necessary to raise the temperature of 1 mole of a substance by 1 °C.

Problem Solving Example:

Q A piece of iron weighing 20 g at a temperature of 95 °C was placed in 100 g of water at 25 °C. Assuming that no heat is lost to the surroundings, what is the resulting temperature of the iron and water? Specific heats: iron = 0.108 cal/g°C; water = 1 cal/g°C.

A The heat lost by the iron must be equal to the heat gained by the water. One solves for the heat lost by the iron by multiplying the number of grams of iron by the number of degrees the temperature dropped by the specific heat of iron. The specific heat of a substance is defined as the amount of heat energy required to raise the temperature of 1 g of a substance by 1 °C. The specific heat for iron is 0.108 cal/g°C. Let t = the final temperature of the system.

amount of heat lost by the iron = 0.108 cal/g°C × 20 g × $(t - 95 \degree C)$

The amount of heat gained by the water is the specific heat of water multiplied by the weight of the water multiplied by the rise in the temperature. The specific heat of water is 1 cal/g°C. Let t = final temperature of the system.

amount of heat gained by water = 1 cal/g°C × 100 g × $(t - 25 \degree C)$

amount of heat lost by the iron $= -$ amount of heat gained by the water

Therefore,

$$(0.108 \text{ cal/g°C}) (20 \text{ g}) (t - 95 \text{ °C}) = -(1 \text{ cal/g°C}) (100 \text{ g}) (t - 25 \text{ °C})$$

$$- 205.2 \text{ cal} + (2.16 \text{ cal/°C})t = -(100 \text{ cal/°C})t + 2,500 \text{ cal}$$

$$2,705.2 \text{ cal} = (102.16 \text{ cal/°C})t$$

$$\frac{2,705.2 \text{ cal}}{102.16 \text{ cal / °C}} = t$$

$$26 \text{ °C} = t.$$

13.2 The First Law of Thermodynamics

The first law of thermodynamics states that the change in internal energy is equal to the difference between the energy supplied to the system as heat and the energy removed from the system as work performed on the surroundings.

$$\Delta E = q - w$$

where E represents the internal energy of the system (the total of all the energy possessed by the system). ΔE is the energy difference between the final and initial states of the system:

$$\Delta E = E_{final} - E_{initial} .$$

The quantity q represents the amount of heat that is added to the system as it passes from the initial to the final state, and w denotes the work done by the system on the surroundings.

Heat added to a system and work done by a system are considered positive quantities (by convention).

For an ideal gas at constant temperature, $\Delta E = 0$ and $q - w = 0$ ($q = w$).

Considering only work due to expansion of a system, against constant external pressure,

$$w = P_{external} \times \Delta V$$

$$\Delta V = V_{final} - V_{initial}$$

Problem Solving Example:

Q You have 1 L of an ideal gas at 0 °C and 10 atm pressure. You allow the gas to expand against a constant external pressure of 1 atm, while the temperature remains constant. Assuming 24.218 cal/L-atm, find q, w, and ΔE in calories.

A The solution of this problem requires a combination of thermodynamics and ideal gas law theory. The gas expands because its pressure is greater than the external pressure. When the pressure of the gas falls to 1 atm, which is the pressure being applied externally, gas expansion will terminate. This allows for determination of the volume change using Boyle's law. You want ΔV (volume change), since $w = \text{work} = P\Delta V$. Boyle's law states that $PV = \text{constant}$ for ideal gas. Thus, $P_1 V_1 = P_2 V_2$. Originally, $P_1 = 10$ atm with $V_1 = 1$ L. You end up with $P_2 = 1$ atm. Thus,

$$V_2 = \frac{P_1 V_1}{P_2} = \frac{(10\,\text{atm})(1\,\text{L})}{(1\,\text{atm})} = 10\,\text{L}.$$

$\Delta V = 10$ L (final) $- 1$ L (original) $= 9$ L. Therefore, $w = P\Delta V = $ (1 atm)(9 L) = 9 L-atm, but there are 24.218 cal per L-atm, so in calories,

$$w = 9 \text{ L-atm} \times 24.218 \text{ cal/L-atm} = 200 \text{ cal}.$$

q = the heat absorbed by the system. To find its value, you employ the first law of thermodynamics, which says $\Delta E = q - w$, where ΔE = change in energy of the system and w = work performed. Since the gas is ideal, E = energy is only a function of temperature. As such, $\Delta E = 0$, since the temperature is constant. You have, therefore, $0 = q - w$, or $q = w$. You just found w, which means q = 200 cal.

13.3 Reversible and Irreversible Processes

In a reversible expansion of a gas, the opposing pressure is virtually equal to the pressure exerted by the gas. It is reversible because any slight increase in the external pressure will reverse the process and cause compression to occur.

The maximum work derived from any change will be obtained only if the process is carried out in a reversible manner.

$$w_{\text{maximum at constant temperature}} = 2.303RT \log \frac{V_2}{V_1}$$

All real, spontaneous changes are therefore not reversible, and the work that can be derived from an irreversible change is always less than the theoretical maximum.

Problem Solving Example:

What is the maximum work, w, that can be derived from the reversible expansion of a gas from a volume of 2.4 L to 5.0 L at a constant temperature of 25 °C?

$w_{\text{max at constant temp}} = 2.303RT \log V_2 / V_1$

$w_{\text{max}} = (2.303)(1.986 \text{ cal/mole K})(298 \text{ K}) \log 5.0/2.4$

$w_{\text{max}} = 430 \text{ cal/mole}$

13.4 Enthalpy

The heat content of a substance is called enthalpy, H. A heat change in a chemical reaction is termed a difference in enthalpy, or ΔH. The term "change in enthalpy" refers to the heat change during a process carried out at a constant pressure:

$$\Delta H = q_p; q_p \text{ means "heat at constant pressure."}$$

The change in enthalpy, ΔH, is defined

$$\Delta H = \Sigma H_{\text{products}} - \Sigma H_{\text{reactants}} .$$

When more than 1 mole of a compound is reacted or formed, the molar enthalpy of the compound is multiplied by the number of moles reacted (or formed).

Enthalpy is a state function. Changes in enthalpy for exothermic and endothermic reactions are shown below:

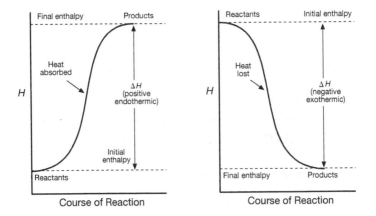

The ΔH of an endothermic reaction is positive, whereas that for an exothermic reaction is negative.

Problem Solving Example:

 Given the following reactions:

$$S(s) + O_2(g) \rightarrow SO_2(g) \qquad \Delta H = -71.0 \text{ kcal}$$

$$SO_2(g) + \frac{1}{2}O_2(g) \rightarrow SO_3(g) \qquad \Delta H = -23.5 \text{ kcal}$$

calculate ΔH for the reaction: $S(s) + 1\frac{1}{2}O_2(g) \rightarrow SO_3(g)$.

A The heat of a given chemical change is the same whether the reaction proceeds in one or several steps; in other words, the energy change is independent of the path taken by the reaction. The heat for a given reaction is the algebraic sum of the heats of any sequence of reactions that will yield the reaction in question. For this problem, one can add the two reactions together to obtain the third. The ΔH is found for the third reaction by adding the ΔH's of the other two.

$$\Delta H = -71.0 \text{ kcal}$$
$$+\Delta H = -71.0 \text{ kcal}$$
$$\overline{\qquad -94.5 \text{ kcal}}$$

$$S(s) + O_2(g) \rightarrow SO_2(g)$$
$$+ \qquad SO_2(g) + \frac{1}{2}O_2(g) \rightarrow SO_3(g)$$
$$\overline{S(s) + O_2(g) + SO_2(g) + \frac{1}{2}O_2(g) \rightarrow SO_2(g) + SO_3(g)}$$

This equation can be simplified by subtracting SO_2 (g) from each side and adding O_2 and $\frac{1}{2}O_2$ together. You have, therefore,

$$S(s) + [O_2(g) + \frac{1}{2}O_2(g)] + SO_2(g) - SO_2(g) \rightarrow SO_3(g) + SO_2(g) - SO_2(g)$$

or
$$S(s) + 1\frac{1}{2}O_2(g) \rightarrow SO_3(g).$$

The ΔH for this reaction is -94.5 kcal, as shown above.

13.5 Heat of Reaction

ΔE is equal to the heat absorbed or evolved by the system under conditions of constant volume:

$$\Delta E = q_v \; (q_v \text{ means "heat at constant volume."}).$$

Since

$$H = E + PV,$$

at constant pressure $\Delta H = \Delta E + P\Delta V$. Note that the term $P\Delta V$ is just the pressure-volume work $((\Delta n)RT)$ for an ideal gas at constant temperature, where Δn is the number of moles of gaseous products minus the number of moles of gaseous reactants. Therefore,

$$\Delta H = \Delta E + \Delta nRT$$

for a reaction that involves gases. If only solid and liquid phases are present, ΔV is very small, so $\Delta H \approx \Delta E$.

Problem Solving Example:

Q The equation for the burning of naphthalene is $C_{10}H_8$ (s) + $2O_2$(g) $\rightarrow$ $10\,CO_2$(g) + $4H_2O$(l). For every mole of $C_{10}H_8$ burned, $-1,226.7$ kcal is evolved at 25 °C in a fixed-volume combustion chamber. $\Delta H°$ for H_2O (l) $= -68.4$ kcal/mole and H_2O (g) $= -57.8$ kcal/mole. Calculate (a) the heat of reaction at constant temperature and (b) ΔH for the case where all of the H_2O is gaseous.

A This problem deals with the heat evolved when a compound is heated with oxygen to form carbon dioxide and water (the heat of combustion). The heat of reaction, or combustion in this case, ΔH, is given by the formula $\Delta H = \Delta E + \Delta nRT$, where $\Delta E =$ amount of heat released per mole, $\Delta n =$ change in moles, $R =$ universal gas constant, and $T =$ temperature in Kelvin (Celsius plus 273). Thus, to answer (a) substitute these values and solve for ΔE. Δn is moles of gas produced $-$

moles of gas reacted, which is 10 moles − 2 moles = 8 moles based on the coefficients in the equation. $T = 25\ °C + 273 = 298$ K. Use R in terms of kilocalories. As such,

$$\Delta H = -\ 1,226.7\ \text{kcal} + (8\ \text{moles})(1.986\ \text{cal/mole K})$$

$$(298\ \text{K})(1\ \text{kcal/1,000 cal})$$

$$= -1,226.7 + 4.7 = -1,222.0\ \text{kcal/mole}.$$

To find (b), note that the reaction is the same, except that H_2O is gaseous, not liquid. You have, therefore, $4H_2O(l) \rightarrow 4H_2O(g)$. The ΔH, change in enthalpy, for this conversion is $4\Delta H°(g) - 4\Delta H°(l)$ of H_2O $= 4(-57.8) - 4(-68.4) = 42.4$ kcal/mole. It follows, then, that the resulting

$$\Delta H = -\ 1,222.0 + 42.4 = -1,179.6\ \text{kcal/mole}.$$

13.6 Hess's Law of Heat Summation

Hess's law of heat summation states that when a reaction can be expressed as the algebraic sum of two or more reactions, the heat of the reaction, $\Delta H_r°$, is the algebraic sum of the heats of the constituent reactions.

The enthalpy changes associated with the reactions that correspond to the formation of a substance from its free elements are called heats of formation, ΔH_f.

$$\Delta H_r° = \Sigma \Delta H_f°(\text{products}) - \Sigma \Delta H_f°(\text{reactants})$$

Problem Solving Example:

Calculate $\Delta H_r°$ for the combustion of methane, CH_4. The balanced reaction is $CH_4(g) + 2O_2(g) \rightarrow CO_2(g) + 2H_2O(l)$.

$\Delta H_f°$ in kcal/mole $= -17.89, 0, -94.05, -68.4$.

ΔH_r° is the standard enthalpy change of reaction. Standard conditions are defined as 25 °C and 1 atm.

Enthalpy is the heat content of a system. If the overall change in enthalpy is negative, then heat is given off to the surroundings and the reaction is called exothermic. When the change is positive, heat is absorbed and the reaction is endothermic. Endothermic compounds are often unstable and can sometimes explode. An endothermic compound, however, is a more efficient fuel because, upon combustion, it yields more heat energy.

ΔH_r° is calculated using the enthalpies of formation, ΔH_f°. The sum of enthalpies of formation of products minus the sum of the enthalpies of formation of reactants, where each product and reactant is multiplied by its molar amount in the reaction as indicated by the coefficients, gives the value of ΔH_r° . In other words,

$$\Delta H_r^\circ = \Sigma \Delta H_f^\circ (\text{products}) - \Sigma \Delta H_f^\circ (\text{reactants}).$$

ΔH_f° for elements is always zero. In this reaction, therefore, ΔH_f° for O_2 is zero.

$$\Sigma \Delta H_f^\circ \text{ products} = -94.05 - 136.8$$

$$\Sigma \Delta H_f^\circ \text{ reactants} = -17.89$$

$$\Delta \Delta H_r^\circ = -94.05 - 136.8 - (-17.89)$$

$$= -213.0 \text{ kcal/mole CH}_4 \text{ burned}$$

This reaction is exothermic, because the ΔH_r° = a negative value.

13.7 Standard States

The standard state corresponds to 25 °C and 1 atm.

Heats of formation of substances in their standard states are indicated as ΔH_f° .

13.8 Bond Energies

For diatomic molecules, the bond dissociation energy, $\Delta H^{\circ}_{\text{diss}}$, is the amount of energy per mole required to break the bond and produce two atoms, with reactants and products being ideal gases in their standard states at 25 °C.

The heat of formation of an atom is defined as the amount of energy required to form 1 mole of gaseous atoms from its element in its common physical state at 25 °C and 1 atm pressure.

In the case of diatomic gaseous molecules of elements, the ΔH°_{f} of an atom is equal to one-half the value of the dissociation energy, that is,

$$-\Delta H^{\circ}_{f} = \frac{1}{2}\Delta H^{\circ}_{\text{diss}}$$

where the minus sign is needed since one process liberates heat, and the other requires heat.

For the reaction HO–OH(g) → 2OH(g),

$$\Delta H^{\circ}_{\text{diss}} = 2\left(\Delta H^{\circ}_{f}\text{OH}\right) - \left(\Delta H^{\circ}_{f}\text{HO} - \text{OH}\right).$$

The energy needed to reduce a gaseous molecule to neutral gaseous atoms, called the atomization energy, is the sum of all of the bond energies in the molecule.

For polyatomic molecules, the average bond energy, $\Delta H^{\circ}_{\text{diss}}$ avg., is the average energy per bond required to dissociate 1 mole of molecules into their constituent atoms.

Problem Solving Example:

Q Given for hydrogen peroxide, H_2O_2:

$$HO - OH(g) \rightarrow 2OH(g) \qquad \Delta H^\circ_{diss} = 51 \text{ kcal/mole.}$$

From this value and the following data calculate: (a) ΔH°_f of OH(g) and (b) C$-$O bond energy; ΔH°_{diss} in $CH_3OH(g)$. ΔH°_f of $H_2O_2(g)$ = -32.58 kcal/mole. ΔH°_f of $CH_3{}^+$ (g) $= 34.0$ kcal/mole. ΔH°_f of $CH_3OH(g) = -47.96$ kcal/mole.

A (a) ΔH°_{diss} for this reaction is equal to the ΔH°_f of the reactants subtracted from the ΔH°_f of the products, where ΔH° equals enthalpy or heat content.

$$\Delta H^\circ_{diss} = 2 \times \Delta H^\circ_f \text{ OH} - \Delta H^\circ_f \text{ HO-OH} = 51 \text{ kcal/mole}$$

$$2 \times \Delta H^\circ_f \text{ OH} = 51 \text{ kcal/mole} + \Delta H^\circ_f \text{ HO-OH}$$

$$\Delta H^\circ_f\text{OH} = \frac{51\,\text{kcal / mole} - 32.58\,\text{kcal / mole}}{2}$$

$$\Delta H^\circ_f\text{OH} = \frac{18.42\,\text{kcal / mole}}{2} = 9.21\,\text{kcal / mole}$$

(b) CH_3OH dissociates by breaking the C$-$O bond:

$$CH_3OH(g) \rightarrow CH_3{}^+(g) + OH^-(g).$$

Therefore, the bond energy of the C$-$O bond equals ΔH°_{diss} equals the sum of the ΔH's of the products minus the ΔH's of the reactants. Thus,

bond energy of C$-$O $= \Delta H^\circ_f$ of $CH_3{}^+(g) + \Delta H^\circ_f$ of OH$^-$(g) $- \Delta H^\circ_f$ of $CH_3OH(g)$

bond energy of C$-$O $= 34.0$ kcal/mole $+ 9.21$ kcal/mole
$+ 47.96$ kcal/mole
$= 91.2$ kcal/mole.

13.9 Spontaneity of Chemical Reactions

In any spontaneous change, the amount of free energy available decreases toward zero as the process proceeds towards equilibrium.

A negative value of ΔG indicates that a reaction can take place spontaneously, but it does not guarantee that the reaction will take place at all.

13.10 Entropy

The degree of randomness of a system is represented by a thermodynamic quantity called the entropy, S. The greater the randomness, the greater the entropy.

A change in entropy or disorder associated with a given system is

$$\Delta S = S_2 - S_1 .$$

The entropy of the universe increases for any spontaneous process:

$$\Delta S_{universe} = (\Delta S_{system} + \Delta S_{surroundings}) \geq 0 .$$

When a process occurs reversibly at constant temperature, the change in entropy, ΔS, is equal to the heat absorbed divided by the absolute temperature at which the change occurs:

$$\Delta S = \frac{q_{reversible}}{T} .$$

13.11 The Second Law of Thermodynamics

The second law of thermodynamics states that in any spontaneous process there is an increase in the entropy of the universe ($\Delta S_{total} > 0$).

$$\Delta S_{\text{universe}} = \Delta S_{\text{total}} = \Delta S_{\text{system}} + \Delta S_{\text{surroundings}}$$

$$\Delta S_{\text{surroundings}} = \frac{-\Delta H_{\text{system}}}{T} \text{ at constants } P \text{ and } T$$

$$T\Delta S_{\text{total}} = -(\Delta H_{\text{system}} - T\Delta S_{\text{system}})$$

The maximum amount of useful work that can be done by any process at constant temperature and pressure is called the change in Gibbs free energy, ΔG:

$$\Delta G = \Delta H - T\Delta S.$$

Another way in which the second law is stated is that in any spontaneous change, the amount of free energy available decreases.

Thus, if $\Delta G = 0$, then the system is at equilibrium.

13.12 Standard Entropies and Free Energies

The entropy of a substance, compared to its entropy in a perfectly crystalline form at absolute zero, is called its absolute entropy, $S°$.

The third law of thermodynamics states that the entropy of any pure, perfect crystal at absolute zero is equal to zero.

The standard free energy of formation, $\Delta G_f°$, of a substance is defined as the change in free energy for the reaction in which 1 mole of a compound is formed from its elements under standard conditions:

$$\Delta G_f° = \Delta H° - T\Delta S_f°,$$

$$\Delta S_f° = \sum \Delta S_{f_{\text{products}}}° - \sum \Delta S_{f_{\text{reactants}}}°$$

and

$$\Delta G_f° = \Delta H° - T\Delta S_f°,$$

Quiz: Chemical Thermodynamics

1. The expression for w, in the first law of thermodynamics, if negative, implies all of the following EXCEPT

    (A) work has been done by the system.

    (B) the total internal energy has decreased.

    (C) a negative amount of work has been done on the system.

    (D) the system has lost heat.

    (E) work has been done on the outside world.

2. The following reaction coordinates cannot be associated with

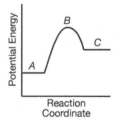

    (A) an endothermic reaction from A to C.

    (B) an exothermic reaction from A to C.

    (C) the activation energy for the reaction.

    (D) the energy for the intermediate.

    (E) the energy absorbed in the reaction from A to C.

3. In the reaction, $CaCO_3(s) \rightarrow CaO(s) + CO_2(g)$ at 95 °C and CO_2 pressure of 1 atm, the ΔH is found to be 176 kJ/mol. Assuming that the volume of the solid phase changes little by comparison with the volume of gas generated, calculate the ΔE in kJ for this reaction. Hint: $\Delta E = \Delta H - P\Delta V$.

   (A) 166   (D) 609

   (B) 245   (E) 413

   (C) 102

4. For the reaction

   $$2HN_3 + 2NO \rightarrow H_2O_2 + 4N_2$$

   with the following molar enthalpies (at 25 °C):

   $HN_3 - H_m° = +264.0$ kJ/mol

   $N_2 - H_m° = 0$ kJ/mol

   $H_2O_2 - H_m° = -187.8$ kJ/mol

   $NO - H_m° = +90.25$ kJ/mol,

   the change in the standard enthalpy for the reaction is

   (A) -896.3 kJ.   (D) $+742.6$ kJ.

   (B) $+937.4$ kJ.   (E) None of the above.

   (C) -309.5 kJ.

5. Using the following table of bond energies, calculate the energy change in the equation.

Bond Energies (in kcal per mole)

H–H	104	C–O	83
H–F	135	C=O	178
H–Cl	103	C–Cl	79
H–Br	88	C–F	105
H–I	71	Si–O	106
Li–H	58	Si–F	136
Cl–Cl	58	C–C	83
C–H	87	C=C	146
O–H	111	C≡C	199
O=O	118	N≡N	225
P–Cl	78		

$CH_4(g) + 2O_2(g) \rightarrow CO_2(g) + 2H_2O(g)$

(A) -111 kcal   (D) -216 kcal

(B) -178 kcal   (E) -145 kcal

(C) $+200$ kcal

6. A reaction is more thermodynamically favored or spontaneous if

(A) ΔH is positive.

(B) ΔS is positive.

(C) ΔH is negative and ΔS is positive.

(D) ΔH is positive and ΔS is negative.

(E) $\Delta H = 0$ and $\Delta S = 0$.

7. A reaction is said to be at equilibrium when

    (A) $\Delta H = 0$.   (D) $\Delta H < 0$.

    (B) $\Delta G = 0$.   (E) $\Delta H > 0$.

    (C) $\Delta S = 0$.

8. For the reaction

$$N_2(g) + O_2(g) \rightarrow 2NO(g)$$

$\Delta G°$ for the reaction is $+ 174$ kJ. This reaction would

    (A) be at equilibrium.

    (B) proceed spontaneously.

    (C) have a negative $\Delta H°$.

    (D) not proceed spontaneously.

    (E) be exothermic.

9. ΔG for a nonspontaneous reaction is

    (A) negative.   (D) small.

    (B) zero.   (E) large.

    (C) positive.

10. The change in ΔS when ice is melted to water is

    (A) 0.

    (B) positive.

    (C) negative.

    (D) Cannot tell from the information given.

    (E) always 0.

ANSWER KEY

1.	(D)	6.	(C)
2.	(B)	7.	(B)
3.	(A)	8.	(D)
4.	(A)	9.	(C)
5.	(D)	10.	(B)

CHAPTER 14

Oxidation and Reduction

14.1 Oxidation and Reduction

Oxidation is defined as a reaction in which atoms or ions undergo an increase in oxidation state. The agent that causes oxidation to occur is called the oxidizing agent and is itself reduced in the process.

Reduction is defined as a reaction in which atoms or ions undergo a decrease in oxidation state. The agent that causes reduction to occur is called the reducing agent and is itself oxidized in the process.

An oxidation number can be defined as the charge that an atom would have if both of the electrons in each bond were assigned to the more electronegative element. The term "oxidation state" is used interchangeably with the term "oxidation number."

The following are the basic rules for assigning oxidation numbers:

1. The oxidation number of any element in its elemental form is 0.

2. The oxidation number of any simple ion (one atom) is equal to the charge on the ion.

3. The sum of all of the oxidation numbers of all of the atoms in a neutral compound is 0.

(More generally, the sum of the oxidation numbers of all of the atoms in a given species is equal to the net charge on that species.)

14.2 Balancing Oxidation-Reduction Reactions Using the Oxidation Number Method

14.2.1 The Oxidation-Number-Change Method

1. Assign oxidation numbers to each atom in the equation.

2. Note which atoms change oxidation number, and calculate the number of electrons transferred, per atom, during oxidation and reduction.

3. When more than one atom of an element that changes oxidation number is present in a formula, calculate the number of electrons transferred per formula unit.

4. Make the number of electrons gained equal to the number lost.

5. Once the coefficients from step 4 have been obtained, the remainder of the equation is balanced by inspection, adding H^+ (in acid solution), OH^- (in basic solution), and H_2O, as required.

Problem Solving Example:

Consider the following balanced equation:

$$Zn + H_2SO_4 \rightarrow ZnSO_4 + H_2 .$$

How does zinc's oxidation number change in this reaction?

A Zinc lacks an oxidation number initially in an uncombined state. It then loses two electrons and becomes $+2$ to couple with the -2 sulfate polyatomic ion in $ZnSO_4$.

14.3 Balancing Redox Equations: The Ion-Electron Method

14.3.1 The Ion-Electron Method

1. Determine which of the substances present are involved in the oxidation-reduction.

2. Break the overall reaction into two half-reactions, one for the oxidation step and one for the reduction step.

3. Balance for mass (i.e., make sure there is the same number of each kind of atom on each side of the equation) for all species except H and O.

4. Add H^+ and H_2O as required (in acidic solutions), or OH^- and H_2O as required (in basic solutions), to balance O first, then H.

5. Balance these reactions electrically by adding electrons to either side so that the total electric charge is the same on the left and right sides.

6. Multiply the two balanced half-reactions by the appropriate factors so that the same number of electrons is transferred in each.

7. Add these half-reactions to obtain the balanced overall reaction. (The electrons should cancel from both sides of the final equation.)

Problem Solving Example:

 Balance the equation for the following reaction taking place in aqueous acid solution:

$$Cr_2O_7^{2-} + I_2 \rightarrow Cr^{3+} + IO_3^-.$$

A The equation in this problem involves both an oxidation and a reduction reaction. It can be balanced by using the following rules: (1) Separate the net reaction into its two major components, the oxidation process (the loss of electrons) and the reduction process (the gain of electrons). For each of these reactions, balance the charges by adding H^+, if the reaction is occurring in an acidic medium, or OH^- in a base medium. (2) Balance the oxygen atoms by adding H_2O. (3) Balance the hydrogen atoms by adding H^+. (4) Combine the two half-reactions, so that all charges from electron transfer cancel out. These rules are applied in the following example.

The net reaction is

$$Cr_2O_7^{2-} + I_2 \rightarrow Cr^{3+} + IO_3^-.$$

The oxidation reaction is

$$I_2 \rightarrow 2IO_3^- + 10e^-.$$

The I atom went from an oxidation number of 0 in I_2 to +5 in IO_3^-, because O always has a −2 charge. You begin with I_2; therefore, 2 moles of IO_3^- must be produced and 10 electrons are lost, 5 from each I atom. Recall that the next step is to balance the charges. The right side has a total of 12 negative charges. Add 12 H^+'s to obtain

$$I_2 \rightarrow 2IO_3^- + 10e^- + 12H^+.$$

To balance the oxygen atoms, add $6H_2O$ to the left side, since there are six O's on the right; thus,

$$I_2 + 6H_2O \rightarrow 2IO_3^- + 10e^- + 12H^+.$$

Hydrogens are already balanced. There are 12 on each side. Proceed to the reduction reaction:

$$Cr_2O_7^{2-} + 6e^- \rightarrow 2Cr^{3+}.$$

Cr began with an oxidation state of $+6$ and went to $+3$. Since $2Cr^{3+}$ are produced, and you began with $Cr_2O_7^{2-}$, a total of 6 electrons are added to the left. Balancing charges: the left side has 8 negative charges and the right side has 6 positive charges. If you add $14H^+$ to the left, they balance. Both sides now have a net $+3$ charge. The equation can now be written:

$$Cr_2O_7^{2-} + 6e^- + 14H^+ \rightarrow 2Cr^{3+}.$$

To balance oxygen atoms, add $7H_2O$'s to the right. You obtain

$$Cr_2O_7^{2-} + 6e^- + 14H^+ \rightarrow 2Cr^{3+} + 7H_2O.$$

The hydrogens are also balanced, 14 on each side. The oxidation reaction becomes

$$I_2 + 6H_2O \rightarrow 2IO_3^- + 10e^- + 12H^+.$$

The reduction reaction is

$$Cr_2O_7^{2-} + 6e^- + 14H^+ \rightarrow 2Cr^{3+} + 7H_2O.$$

Combine these two in such a manner that the number of electrons used in the oxidation reaction is equal to the number used in the reduction. To do this, note that the oxidation reaction has $10e^-$ and the reduction $6e^-$. Both are a multiple of 30. Multiply the oxidation reaction by 3 and the reduction reaction by 5, obtaining oxidation: $3I_2 + 18H_2O \rightarrow 6IO_3^- + 30e^- + 36H^+$

reduction: $5CrO_7^{2-} + 30e^- + 70H^+ \rightarrow 10Cr^{3+} + 35H_2O.$

Add these two half-reactions together.

$$3I_2 + 18H_2O \rightarrow 6IO_3^- + 30e^- + 36H^+$$

$+ \qquad 5Cr_2O_7^{2-} + 30e^- + 70H^+ \rightarrow 10Cr^{3+} + 35H_2O$

$3I_2 + 18H_2O + 5Cr_2O_7^{2-} + 30e^- + 70H^+ \rightarrow 10Cr^{3+} + 35H_2O + 30e^- + 36H^+ + 6IO_3^-$

Simplifying, you obtain:

$$3I_2 + 5Cr_2O_7^{2-} + 34H^+ \rightarrow 6IO_3^- + 10Cr^{3+} + 17H_2O.$$

This is the balanced equation.

14.4 Non-Standard-State Cell Potentials

For a cell at concentrations and conditions other than standard, a potential can be calculated using the following Nernst equation:

$$E_{cell} = E_{cell}^\circ - \frac{0.059}{n} \log Q$$

where E_{cell}° is the standard-state cell voltage, n is the number of electrons exchanged in the equation for the reaction, and Q is the mass action quotient (which is similar in form to an equilibrium constant).

For the cell reaction,

$Zn + Cu^{2+} \rightarrow Cu + Zn^{2+}$, the terms $Q = \left[Zn^{2+} \right] / \left[Cu^{2+} \right]$

the Nernst equation takes the form:

$$E = E^\circ - \frac{0.059}{n} \log \frac{\left[Zn^{2+} \right]}{\left[Cu^{2+} \right]}.$$

Problem Solving Example:

 Given $Zn \rightarrow Zn^{2+} + 2e^-$ with $E^\circ = +0.763$ V, calculate E for a Zn electrode in which $Zn^{2+} = 0.025$ M.

 This problem calls for the use of the Nernst equation, which relates the effect of concentration of ions in a cell on the voltage. It is stated

$$E = E^\circ - \frac{0.059}{n} \log Q, \text{ at } 25\,^\circ C$$

where E = potential under conditions other than standard, E° = standard electrode potential, n = number of electrons gained or lost in the reaction, and Q = ratio of concentration of products to reactants. You are given E° and $n = 2$, since Zn loses two electrons to become Zn^{2+}. For this problem,

$$Q = \left[Zn^{2+} \right].$$

You are given that $[Zn^{2+}] = 0.025$ M.

To find E, substitute these values into the Nernst equation.

$$E = 0.763 - \frac{0.059}{2} \log 0.025$$
$$= 0.763 - \left[(0.0295)(-1.602) \right]$$
$$= 0.810 \text{ V}$$

14.5 Electrolytic Cells

Reactions that do not occur spontaneously can be forced to take place by supplying energy with an external current. These reactions are called electrolytic reactions.

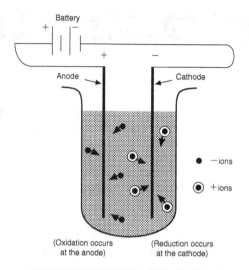

(Oxidation occurs
at the anode)   (Reduction occurs
at the cathode)

ELECTROCHEMICAL REACTIONS

In electrolytic cells, electrical energy is converted into chemical energy.

14.6 Faraday's Law

One faraday is 1 mole of electrons.

(1 F = 1 mole of electrons (F is the "Faraday").

1 F ~ 96,500 coulombs; the charge 1 F is approximately 96,500 coulombs.

One coulomb is the amount of charge that moves past any given point in a circuit when a current of 1 ampere (amp) is supplied for 1 second. (Alternatively, 1 ampere is equivalent to 1 coulomb/second.)

Faraday's law states that during electrolysis, the passage of 1 faraday through the circuit brings about the oxidation of one equivalent weight of a substance at one electrode (anode) and reduction of one equivalent weight at the other electrode (cathode). Note that in all cells, oxidation occurs at the anode and reduction occurs at the cathode.

Problem Solving Example:

Q You want to plate out 50 g of copper from an aqueous solution of $CuSO_4$. Assuming 100% efficiency, how many coulombs are required?

A In this question, you are dealing with the phenomenon of electrolysis. When an electric current is applied to a solution containing ions, the ions will either be reduced or oxidized to their electronically neutral state.

To answer this question, you must realize that Cu^{2+} ions exist in solution. To plate out copper, two electrons must be added to obtain the copper atom, Cu^0. Since the Cu^{2+} must gain electrons, it must be reduced. The amount of electricity that produces a specific amount of reduction (or oxidation) is related by $q = nF$ (Faraday's law), where q = the quantity of electricity in coulombs, n = the number of equivalents oxidized or reduced, and F = faradays. The number of equivalents equals the equivalent weight of the material (M_{eq}), i.e.,

$$N = \frac{m}{M_{eq}}.$$

A faraday = 96,500 coulombs or 1 mole of electrons.

Since a copper ion requires two electrons for reduction, the gram-equivalent weight is one-half of the atomic weight, or 31.8 g-equiv. You have, therefore,

$$q = \frac{(50)}{(31.8)}(96,500) = 1.52 \times 10^5 \, C$$

required.

14.7 Voltaic Cells

One of the most common voltaic cells is the ordinary "dry cell" used in flashlights. It is shown in the drawing below, along with the reactions occurring during the cell's discharge.

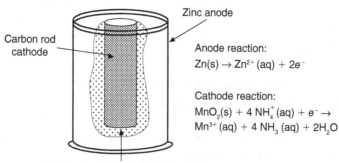

Carbon rod cathode

Zinc anode

Anode reaction:
$Zn(s) \rightarrow Zn^{2+}(aq) + 2e^-$

Cathode reaction:
$MnO_2(s) + 4 NH_4^+(aq) + e^- \rightarrow$
$Mn^{3+}(aq) + 4 NH_3(aq) + 2H_2O$

Moist paste of MnO_2, C, NH_4Cl, H_2O

In galvanic or voltaic cells, the chemical energy is converted into electrical energy.

In galvanic cells, the anode is negative and the cathode is positive (the opposite is true in electrolytic cells).

The force with which the electrons flow from the negative electrode to the positive electrode through an external wire is called the electromotive force, or emf, and is measured in volts (V):

$$1V = \frac{1J}{C}.$$

The greater the tendency or potential of the two half-reactions to occur spontaneously, the greater will be the emf of the cell. The emf of the cell is also called the cell potential, E_{cell}. The cell potential for the Zn/Cu cell can be written

$E°_{cell} = E°_{Cu} - E°_{Zn}$, where the $E°$'s are standard reduction potentials.

The overall standard cell potential is obtained by subtracting the smaller reduction potential from the larger one. A positive emf corresponds to a negative ΔG and therefore to a spontaneous process.

$$\Delta G = -nFE$$

Also,

$$E = E^\circ - \frac{RT}{nF}\ln Q$$

which is the Nernst equation.

$$E = E^\circ - \frac{0.059}{n}\log Q$$

and is analogous to

$$\Delta G = \Delta G^\circ + RT\ln Q.$$

Problem Solving Example:

 Calculate E°, ΔG°, and K for the following reaction at 25 °C:

$$Cu(s) + Cl_2(g) \rightarrow Cu^{2+} + 2Cl^-.$$

A The magnitude of the standard electrode potential, E°, is a measure of the tendency of the half-reaction to occur in the direction of reduction. The standard emf (electromotive force) of a cell is equal to the standard electrode potential of the cathode minus the standard electrode potential of the anode,

$$E^\circ = E^\circ_{cathode} - E^\circ_{anode}.$$

The reaction taking place at the anode is written as an oxidation reaction, and the reaction taking place at the cathode is written as a reduction reaction. The cell reaction is the sum of these two reactions. Thus,

$$\text{Oxidation: } Cu(s) \rightarrow Cu^{2+} + 2e^-$$

$$\text{Reduction: } Cl_2(g) + 2e^- \rightarrow 2Cl^-$$

Since the oxidation reaction is to occur at the anode and the reduction at the cathode, the cell is written as

$$Cu|Cu^{2+}\|Cl^-\|Cl_2(g)|Pt$$

$$E^\circ = E^\circ_{Cl^-|Cl_2|Pt} - E^\circ_{Cu^{2+}|Cu}.$$

E° can be found on any Standard Electrode Potential Table.

$$E^\circ_{Cl^-|Cl_2|Pt} = 1.360 \text{ V} \qquad E^\circ_{Cu^{2+}|Cu} = 0.337 \text{ V}$$

Thus, $E^\circ = 1.360 \text{ V} - 0.337 \text{ V}$

$$E^\circ = 1.023 \text{ V}.$$

Now that E° is known, standard Gibbs free energy (ΔG°), i.e., the energy available to do useful work, can be calculated from $\Delta G^\circ = -nF E^\circ$, where n = number of electrons transferred and F = value of a faraday of electricity (23,060 cal/V-mole). From the redox half-reactions written, one sees that two electrons are being transferred. Consequently, $n = 2$.

Substituting these values, one obtains

$$\Delta G^\circ = -\left(2 \text{ moles}\right)\left(23,060\frac{cal}{V\text{-}mole}\right)\left(1.023 \text{ V}\right)$$

$$= -47,180 \text{ cal}.$$

The equilibrium constant, K, can be calculated using the Nernst equation for unit activity. In other words, $E° = (0.059/n)$ log K. Substituting the previously calculated values for $E°$ and n,

$$1.023 \text{ V} = (0.059/2) \log K.$$

Solving, log $K = 34.68$, or $K = 4.764 \times 10^{34}$.

Quiz: Oxidation and Reduction

1. Which of the following correctly describes a change in oxidation state for the reaction $H_2 + S \rightarrow H_2S$?

    (A) $H_2 + 2e^- \rightarrow 2H^-$

    (B) $H_2 \rightarrow 2H$

    (C) $S + 2e^- \rightarrow S^{2-}$

    (D) $S^{2-} \rightarrow S + 2e^-$

    (E) None of the above.

2. A strip of zinc is dipped in a solution of copper (II) sulfate. Select the correct occurring half-reaction.

    (A) $Cu^{++} + 2$ electrons $\rightarrow Cu$, reduction

    (B) $Cu + 2$ electrons $\rightarrow Cu^{++}$, reduction

    (C) $Cu \rightarrow Cu^{++} + 2$ electrons, reduction

    (D) $Zn \rightarrow Zn^{++} + 2$ electrons, oxidation

    (E) $Zn + 2$ electrons $\rightarrow Zn^{++}$, oxidation

3. In the reaction

    $$Zn(s) + 2HCl(aq) \rightarrow ZnCl_2(aq) + H_2(g)$$

    (A) zinc is oxidized.
    (B) the oxidation number of chlorine remains unchanged.
    (C) the oxidation number of hydrogen changes from $+1$ to 0.
    (D) both (A) and (B).
    (E) (A), (B), and (C).

4. In the following unbalanced reaction, which reactant acts as an oxidizing agent?

    $$HClO + Sn^{2+} + H^+ \rightarrow Cl^- + Sn^{4+} + H_2O$$

    (A) H^+ (D) $HClO$

    (B) Sn^{4+} (E) Cl^-

    (C) Sn^{2+}

5. Consider the following balanced equation:

    $$2K + 2HCl \rightarrow 2KCl + H_2 .$$

    The respective oxidation numbers for K, H, and Cl before and after the reaction

    (A) go from $0, -1, +1$ to $-1, +1, 0$.
    (B) go from $0, +1, -1$ to $+1, -1, 0$.
    (C) go from $1, -1, 0$ to $-1, +1, -1$.
    (D) go from $1, -1, 0$ to $-1, -1, 0$.
    (E) go from $0, 0, 1$ to $1, 1, -1$.

6. The reaction $2H^+(aq) + 2e^- \rightarrow H_2(g)$ is an example of

    (A) an oxidation.

    (B) a reduction.

    (C) an oxidation-reduction.

    (D) the reaction at the hydrogen anode.

    (E) an addition reaction.

7. The site of oxidation in an electrochemical cell is

    (A) the anode.   (D) the salt bridge.

    (B) the cathode.   (E) none of the above.

    (C) the electrode.

8. One Faraday of electricity is passed through an HCl electrolyte solution. Select the correct electrode result.

    (A) 1 gram of chloride ions is deposited at the anode.

    (B) 1 gram of hydrogen ions is deposited at the cathode.

    (C) 5 grams of hydrogen ions are deposited at the anode.

    (D) 35 grams of chloride ions are deposited at the anode.

    (E) 36.5 grams of chloride ions are deposited at the cathode.

9. Which of the following equations can be used to calculate the emf of a voltaic cell at various concentrations?

    (A) $E = E° \dfrac{-0.059}{n} \log Q$

    (B) $E = q - w$

    (C) $E = E°_{products} - E°_{reactants}$

    (D) $E = E° \dfrac{-0.059}{n} \ln Q$

    (E) None of the above.

10. What is the potential of a half cell consisting of a platinum wire dipped into a solution 0.01 M in Sn^{2+} and 0.001 M in Sn^{4+} at 25 °C?

    (A) $E°_{oxid} + 0.059$ (D) $E°_{oxid} - 0.059$

    (B) $E°_{red} - \dfrac{0.059}{2}$ (E) $E°_{red}$

    (C) $E°_{red} + \dfrac{0.059}{2}$

ANSWER KEY			
1.	(C)	6.	(B)
2.	(D)	7.	(A)
3.	(E)	8.	(B)
4.	(D)	9.	(A)
5.	(B)	10.	(B)

The Periodic Table

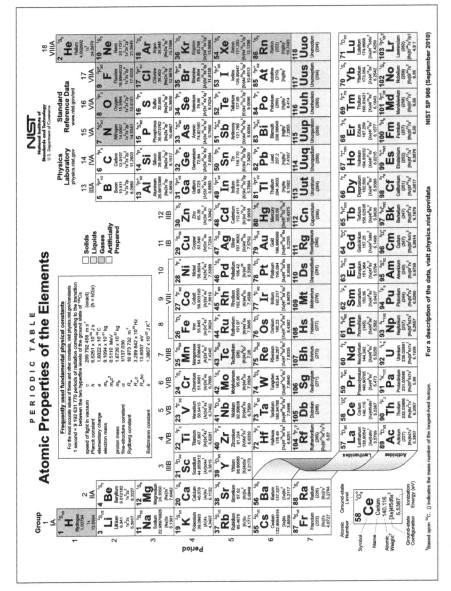

Source: *National Institute of Standards and Technology*

NOTES

NOTES

NOTES

NOTES

NOTES

NOTES

NOTES

NOTES

NOTES

NOTES

REA's Study Guides

Review Books, Refreshers, and Comprehensive References

Problem Solvers®

Presenting an answer to the pressing need for easy-to-understand and up-to-date study guides detailing the wide world of mathematics and science.

High School Tutors®

In-depth guides that cover the length and breadth of the science and math subjects taught in high schools nationwide.

Essentials®

An insightful series of more useful, more practical, and more informative references comprehensively covering more than 150 subjects.

Super Reviews®

Don't miss a thing! Review it all thoroughly with this series of complete subject references at an affordable price.

Interactive Flashcard Books®

Flip through these essential, interactive study aids that go far beyond ordinary flashcards.

Reference

Explore dozens of clearly written, practical guides covering a wide scope of subjects from business to engineering to languages and many more.

For information about any of REA's books, visit
www.rea.com

Research & Education Association
61 Ethel Road W., Piscataway, NJ 08854
Phone: (732) 819-8880